# Praise for *The Lean M*

"Oftentimes, the desire in farming is to open up more land, grow more crops, and get bigger. In *The Lean Micro Farm*, Ben Hartman doesn't just illuminate the enormous potential in *getting small*—for communities, for the environment, for the profitability of farms— he lays out a roadmap for how to achieve it.

"As Ben eloquently demonstrates, small doesn't mean less, small can just as easily mean more. Small can mean better. Small can mean, in the immortal words of economist E. F. Schumacher, *beautiful*. More importantly, when the desire is to grow more and earn more, sometimes getting smaller is actually the answer.

"I was delighted and a bit terrified to pick up Ben Hartman's new book, because every time I read something Ben wrote, significant portions of my farm change. And *The Lean Micro Farm* is no exception. Chapter by chapter you see the ways in which shrinking their farm has led Ben and his wife Rachel to a happier, healthier, more sustainable, more localized farm without risking income. Each section is filled with examples and strategies for how they got small and what it looks like in practice. It's well-written, thought-provoking, and potentially life-altering. I immediately found myself penciling out ways to make our farm smaller.

"So fair warning, this book will change your farm."

—JESSE FROST, author of
*The Living Soil Handbook*

"*The Lean Micro Farm* is a game changer for farming and food production. With well-thought-out principles and innovative techniques for planning and maintaining profitable tiny farms, Ben Hartman opens the door to a future of micro farms everywhere, rather than fewer and fewer large farms in rural locations only. This easy-to-read book is full of time-saving and ecologically sustainable techniques, such as flipping beds of both short and tall-growing crops with minimal soil disturbance so that multiple crops can be grown well each year in a small space. Ben's tested methods can be applied to gardens and homesteads as well as small farms. Thank you, Ben, for bringing the ideas of my hero, E. F. Schumacher, into the 21st century and showing that they are as relevant as they were when his book, *Small Is Beautiful*, was first published!"

—HELEN ATTHOWE, Woodleaf Farm,
Montana; author of *The Ecological Farm*

"Ben is a shining example of the powerful ideas and efficient methods he describes. He has a way of making things simple and a simple way of explaining them! Small is beautiful and small makes sense, now more than ever. Ben's one third of an acre is understandable, achievable, and hugely productive of nutritious food. It's my pleasure to learn more about and endorse his approach. Here's to health with Hartman."

—CHARLES DOWDING, author of
*No Dig Gardening, No Dig Cookbook*,
and *No Dig Children's Gardening Book*

"In the field, Ben never zags. The crops are in perfect alignment. Zagging would be wasteful motion. However, in a world that's obsessed with scaling, whether the business is technology or 'never-enough farming,' Ben has zagged by getting small. For the sake of his family and community, Ben simply wants to live better and work less. Don't we all? In this book, he explains how to achieve that goal by getting small with lean thinking. In other words, he explains how the philosophy of 'just enough' is a zag we should all consider putting into practice."

—JOSH HOWELL, president and executive
team leader, Lean Enterprise Institute

# Also by Ben Hartman

*The Lean Farm:*
*How to Minimize Waste,*
*Increase Efficiency, and Maximize*
*Value and Profits with Less Work*

*The Lean Farm Guide to Growing Vegetables:*
*More In-Depth Lean Techniques*
*for Efficient Organic Production*

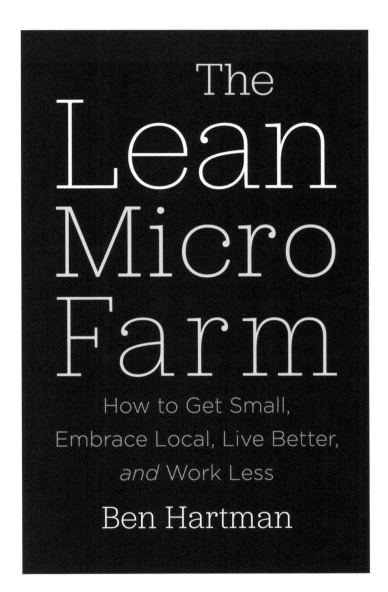

# The Lean Micro Farm

## How to Get Small, Embrace Local, Live Better, *and* Work Less

### Ben Hartman

Chelsea Green Publishing
White River Junction, Vermont
London, UK

Project Manager: Rebecca Springer
Developmental Editor: Ben Trollinger
Copy Editor: Hope Clarke
Proofreader: Angela Boyle
Indexer: WordCo Indexing Services, Inc.
Designer: Melissa Jacobson
Page Layout: Abrah Griggs

Printed in the United States of America.
First printing November 2023.
10 9 8 7 6 5 4 3 2 1      23 24 25 26 27

**Our Commitment to Green Publishing**
Chelsea Green sees publishing as a tool for cultural change and ecological stewardship. We strive to align our book manufacturing practices with our editorial mission and to reduce the impact of our business enterprise in the environment. We print our books using vegetable-based inks whenever possible. This book may cost slightly more because it was printed on paper from responsibly managed forests, and we hope you'll agree that it's worth it. *The Lean Micro Farm* was printed on paper supplied by Versa that is certified by the Forest Stewardship Council.®

**Library of Congress Cataloging-in-Publication Data**
Names: Hartman, Ben, 1978– author.
Title: The lean micro farm : how to get small, embrace local, live better, and work less / Ben Hartman.
Description: First edition | White River Junction, Vermont : Chelsea Green Publishing, November 2023 | Includes bibliographical references and index. |
Identifiers: LCCN 2023034586 | ISBN 9781645022046 (paperback) | ISBN 9781645022053 (ebook)
Subjects: LCSH: Organic gardening. | Gardening. | Sustainable agriculture. | Handbooks and manuals. | BISAC: TECHNOLOGY & ENGINEERING / Agriculture / Sustainable Agriculture | TECHNOLOGY & ENGINEERING / Agriculture / Organic
Classification: LCC SB454 .H348 2023 | DDC 635/.0484—dc23/eng/20230825
LC record available at https://lccn.loc.gov/2023034586

Chelsea Green Publishing
White River Junction, Vermont, USA
London, UK

www.chelseagreen.com

# Contents

Original artwork by Rachel Shenk.

# Acknowledgments

I want to thank the farm crew at Clay Bottom Farm, who tended our gardens and delivered our food while I wrote this book. Specifically, thank you to Andrew Ness, Sierra Ross Richer, and Christian Berambaye. A special thank-you to Nicole Craig, assistant farm manager at Clay Bottom Farm, who arranged interviews, performed research, and edited portions of this book. Without help from each of these people, this book would not exist.

My friend Jim Womack, founder and chairman of the Lean Enterprise Institute, has been a source of gentle encouragement and insight over the course of many years. Jim's passion for banishing waste—for doing better with less—is infectious. His teaching inspired me to innovate many of the approaches that appear in this book. Thank you, Jim.

Jon Byler, my neighbor, good friend, and farming colleague, died during the writing of this book. Jon donated cement slabs and electrical parts for our greenhouses, helped me fix sprinklers and skid loader tires, and happily pulled out his welder whenever I asked. He heavily influenced the design and final shape of our farm and thus this book. I miss you, Jon. Thank you.

Many exemplary microscale farmers allowed me to interview them for this book. Thank you especially to Gregory Alan Isakov, Glory Adonase, Ben Brown, Julia Whitney Brown, Deijhon Yearby, Tamara Bogolasky, and Attila Szocs. Thanks to my colleagues at Winrock International and USAID Nigeria for your enthusiastic support of lean farming and for arranging the interview with Adonase. Specifically, thank you to Ben Odoemena, PhD, chief of party for Feed the Future Nigeria Agricultural Extension and Advisory Services Activity; Jean-Pierre Rousseau, associate director of agriculture, resilience, and water at Winrock International; Charity Hanif, Winrock agribusiness consultant; and Chyka Okarter, value chain advisor for Feed the Future Nigeria Agricultural Extension and Advisory Services Activity.

Shannon McDevitt, a student in the Moreau College Initiative at Westville Correctional Facility, transcribed my interview with Szocs, the Romanian plum farmer I write about in chapter 5. Thank you, Shannon. Thank you to Dr. Justus Ghormley, assistant director of college operations at the Moreau College Initiative, for facilitating the work between Shannon and me.

Thank you to Goshen, Indiana, artist Rachel Shenk, who created the woodblock print on page vi, and to Gregory Lehman, a builder and artist in Goshen, who drew the renderings of our barn-house found on page 247. Thank you to Darrell Ness for reading an early draft and providing comments.

Thanks also to Ben Trollinger, my editor at Chelsea Green, who provided valuable insight at key moments, and to the Chelsea Green production team, specifically Rebecca Springer, Melissa Jacobson, and Jenna Cunniff, who turned my manuscript into the beautiful book in your hands.

Finally, a special thank-you to my wife, Rachel, for encouraging me to write this book and for tending to our kids and to myriad other duties while I did so. Rachel also edited this manuscript. Thanks to Arlo and Leander, our children, for farming and playing by our side, and for pulling us away from our work. I am proud to be part of this family.

# Introduction

*Get big or get out.*

—EARL BUTZ, US Secretary of Agriculture, 1971–1976

*Get small and stay in.*

—GENE LOGSDON, small farmer and author of
*The Contrary Farmer*

In 2017, my wife, Rachel, and I moved our rural vegetable farm into the city primarily to be closer to schools, extended family, and activities for our two young kids.

The move also allowed the farm, then 11 years old, a fresh start. We microsized from 1 acre down to ⅓ acre in production. Instead of seven buildings, we built one—a barn-house. Instead of four greenhouses, we put up two. Our inspirations included the connected farms of Maine and New Hampshire, Edo-period Japanese farms, and the *hof* that my German ancestors used as a house and an animal barn. We now sell all of our food within 1½ miles of the farm. We switched to a no-till deep-mulch system, utilizing 4 inches of compost on the soil surface. Instead of feeding our plants minerals that were mined and shipped from far away, we now rely only on our own compost, made from local leaves, for fertility. We use just seven field tools to complete most work. We still grow a wide range of crops but chose five focus crops; all of our crops are grown within 60 steps of our barn-house.

Yet while our farm is smaller and we work less, we earn as much as before.

To be honest, we weren't always certain, in the first year or two, whether we'd made the right call. To move a business and our family at the same time felt risky, even crazy at times. But in the end, we're grateful for our move and lifestyle redesign. We have more time for our kids, and we are more deeply rooted in our community.

Not every day is perfect. To do it again, there's plenty we would change. But this pivot to a microscale farm allowed us a new beginning. This book tells our get-small story. I will explain as plainly as I can how we earn a comfortable living on our tiny farm and show how you can do the same.

Rachel in the garden with Leander. When our kids were born, our priorities shifted.

## Inspiration to Get Small

Our first son, Arlo, was born in 2014, followed by Leander in 2016. At the time of their births, we had just finished a "lean transformation" on our farm. A few years prior, Steve Brenneman, a CSA customer and founder of an aluminum manufacturing business, had come out to our farm and helped us streamline our operation by applying the Japanese lean-manufacturing system. His ideas helped us root out waste and dramatically increase our productivity and profits in just a few seasons.[1] Our farm was organized and clean rather than cluttered. We grew higher-quality crops with less work. We were finally earning a living wage and selling everything we could grow.

But now, with kids, things were different. We were frustrated, trying to juggle kid duty and farmwork, even with lean systems on our farm. We wanted these years as young parents to be a joyful time, but they were increasingly defined by stress and overwork. We wanted a slower pace with more downtime.

The year Arlo was born, my book *The Lean Farm*, which tells the story of our farm's lean evolution, was published. I was invited to speak at conferences, visit and consult on farms, and partner with development projects. Many farmers from all over the world were inspired by the book to grow their businesses by getting rid of waste. That winter, while Rachel cared for the boys—with generous help from her mom—I took eight trips, during which I visited dozens of inspiring high-profit, tiny-scale farms.

Over the next few years, I wrote a growing manual to accompany *The Lean Farm*, and I traveled more. Every place I visited gave me ideas to incorporate into our farm as I pondered what our future farm and life might look like.

In Vancouver, British Columbia, I visited Doug Zaklan and Gemma McNeil at Zaklan Heritage Farm. On a mere 1½ acres of land, nestled between apartment complexes, the two young farmers were making a living and providing good work for several employees by carefully setting limits. They chose their accounts—in this case, restaurants—carefully, cultivating deep, long-lasting relationships with the chefs they sold their food to.

While on a tour in England, I stayed with Charles Dowding, the gardener, educator, and author of several books on "no-dig" gardening. Though I arrived at Dowding's Homestead Acres after dusk, he grabbed a flashlight and excitedly showed me his immaculate home garden, with beds recently covered with the annual 1 inch of new compost. The edges of his plots, between soil and grass, were perfectly cut lines. Dowding had simplified everything, especially in how he'd prepared soils. "People seem to love loose soil. But a lot of gardeners overwork their land," he told me. "A soil's natural condition is firm. Look at the grass we are standing on," he exclaimed as he pounded the ground with his foot. "It's growing just fine, and the ground is firm!"

At Trill Farm Garden, in Axminster, UK, Ashley Wheeler and Kate Norman showed me their approach to farming with friends on a rented piece of ground. Some in the community worked full-time on the farm while others worked off-farm jobs, and they came together frequently to laugh and share food. I was inspired to see these young farmers using creative scheduling and collaboration to support each other's work.

On a trip to South Wales, I stayed with Tom O'Kane at Cae Tân Farm. O'Kane comanages Cae Tân with Eva Jones, Ruth Evans, and a group of volunteer directors. The farmers at Cae Tân have designed an ownership model that deeply integrates their

Committee members and volunteers sharing a meal at Cae Tân Farm in Wales. Photo courtesy of Cae Tân Farm.

community. A nonprofit committee, made up of customers and friends, actually owns the assets of the farm. Large purchasing decisions have to be approved by the nonprofit. Committee members and volunteers visit the farm frequently because they are invested in it. Cae Tân uses the motto "Our members and volunteers shape us." The local community, for Cae Tân, is pulled to the center of the farm.

I found a common theme among these growers: they used constraint as a strength, not a weakness. Their farms were exemplary not in spite of being small but because of being small.

# Simplicity + Constraint: Principles for Getting Smaller

As we discussed my travels, Rachel and I chose five principles to guide us as we set up our new micro farm:

1. Leverage constraints.
2. Build *just enough* (then maximize fixed costs).
3. Essentialize.
4. Simplify fieldwork.
5. Localize.

These principles help balance the life we want with the work that we are doing. They allow us to farm and live better. These are principles that we try to follow even today, though we are a long way from mastering them. The impulse and incentives to get bigger, do more, and sell food farther away from home are around us all the time. The default mode in our economy is constant growth. Our five principles are like anchors: they ground our decisions against the lure of getting bigger.

More specifically, here is how these ideas shaped our tiny farm.

### 1. LEVERAGE CONSTRAINTS.

Carefully set limits of size, time, and business shape at the outset.

We started by evaluating the scale of farm we wanted, given our new reality of farming with kids. How many hours did we want to devote to the business? How many acres did we need to farm in order to meet our financial goals? How far away should we deliver our food? Carefully chosen boundaries helped us keep our focus during our 12 months of construction, and they continue to guide us even now.

## 2. BUILD *JUST ENOUGH* (THEN MAXIMIZE FIXED COSTS).

Use the right amount of infrastructure: not too much, not too little.

We worked hard to build a just enough farm, where every bit of infrastructure would be well used. We designed our farm taking cues from efficient and small-scale preindustrial farmstead models that relied on multipurpose spaces and creative farm layouts that made the most of every square foot.

## 3. ESSENTIALIZE.

Focus on the most vital crops and accounts.

The next task was to edit, or to separate the vital few from the trivial many, as Vilfredo Pareto, an Italian economist, once put it. According to Pareto, roughly 80 percent of outcomes stem from 20 percent of causes (the "vital

The author with son Arlo. Having kids has helped me appreciate what it means to work well.

few"). Thus, about 20 percent of our time, energy, and effort was likely producing 80 percent of our profits. We used this concept to carefully choose five focus crops and a small handful of accounts that order on a consistent basis. We can count on fairly predictable weekly orders from our grocery and restaurant accounts. This tight focus is a key reason behind our financial stability, and it allows a workload that feels balanced, with enough time for family outside of farming.

## 4. SIMPLIFY FIELDWORK.

Farm with the simplest approach possible, with just a few tools.

Next, we set out to simplify our work. Could we farm with fewer tools? Could we prepare growing beds with fewer steps? Could we farm with less digging, less gas, less fertilizer, and less plastic? Could we wash lettuce with less water? Every process could be simplified. It was just a matter of discovering how.

## 5. LOCALIZE.
Put the community at the center of the farm.

In *The Lean Farm*, I wrote about *genchi gembutsu*, the Japanese product-design practice of listening deeply to customers to thoroughly understand them and what they are willing to pay for. At our new farm, we wanted to add another concept: Gandhi's "spirit of *Swadeshi*," or self-reliance at the village level. How might we design a farm that is deeply woven into the local community?

# Small vs. Lean: Similarities and Differences

The get-small principles in this book are similar to lean principles. They both focus on getting rid of nonessentials to create more productive and satisfying workspaces. They both have the potential to make a business more profitable.

But there is a key difference. Lean tools apply to any size business. In fact, many giant-size businesses, like major airlines and automakers, use lean ideas to get even bigger. With less waste, these businesses can grow more quickly. On the other hand, the get-small principles in this book apply to micro enterprises, like small farms, that want to stay small or get smaller.

Put another way: Lean is a powerful efficiency toolkit. Get-small principles are tools to harness the specific advantages of small. They are a guide for doing less but better.

We've found that *lean* and *small* pair well together. Leaning-up pushed us to appreciate our small size. We wouldn't want to compete with the gigantic mass-production farms selling to Walmart and other big box stores in our area. But if we stay lean and nimble, keeping costs low while focusing on high-value crops, we can carve out a niche selling specialty food directly to customers. Lean and get-small principles, put together, have leveled the playing field, allowing our tiny farm to support our family in the same way that 400 acres of commodity crops supported my family growing up. The difference is that now Rachel and I have the satisfaction of knowing that our food is eaten in our local community.

# Small Is Beautiful on the Farm

The farmers I visited after my lean books were published opened my eyes to possibilities for our future. Another source of ideas came from E. F. Schumacher's *Small Is Beautiful: Economics as If People Mattered*, written in 1972.

# The Lean System Summarized

*Lean* is fundamentally about scouring your workplace for waste, rooting it out, and then replacing the waste with activities that add value. Thus, the formula at the heart of lean: eliminated waste = capacity. With this approach, it is possible to become more profitable while actually doing less. The lean system, as we put it in place on our farm, involves four steps.

1. *Ruthlessly sort.* First, we decluttered and organized the farm with a lean organizational method called 5S. I describe 5S in more detail in chapter 4 of this book. In a nutshell, with the Japanese 5S system, only the items you use frequently should be in production spaces; everything else should be removed.

2. *Precisely identify value.* Next, we interviewed our customers, asking them what they wanted, when they wanted it, and what amount, in order to arrive at a precise understanding of what they valued.

3. *Cut out the waste.* Then, we became detectives. We took a magnifying glass to our production systems, looking for activities that weren't adding value. Over a period of four years, we rooted out 10 types of farm waste:

   - Overproduction
   - Waiting
   - Transportation
   - Overprocessing
   - Excess inventory
   - Motion
   - Defective products
   - Overburdening (*muri*)
   - Uneven production and sales (*mura*)
   - Unused talent

4. *Practice* kaizen, *or continuous improvement.* Finally, we put into place routines to ensure that every season, we organize more thoroughly, deepen our relationship with customers, and cut out more waste. The end result is a more efficient farm with higher profits and less work.

Cutting greens in a high tunnel. Photo courtesy of chuck.studio.

Shishito peppers.

I first encountered *Small Is Beautiful* as a junior in high school in the late 1990s, when I was experimenting with radical simplicity. That year I grew my hair out, got rid of half my clothes, helped start a garden club at school, and started my first CSA as an after-school and summer job. I read and reread Thoreau's ode to simplicity, *Walden*. I saw *Small Is Beautiful* on a library shelf and was drawn in by the title. I grew up on a corn and soybean farm. All around me were big tractors and giant fields. I knew that I wanted to grow food, but I wondered if it could be done profitably on a smaller scale. Schumacher's book convinced me that the answer was a resounding "yes."

When Rachel and I began talking about lifestyle changes we wanted to make after our kids were born, I pulled my tattered copy of *Small Is Beautiful* off the shelf. On the front page, a previous owner of the book had written, "HUMANS ARE SMALL THEREFORE SMALL IS BEAUTIFUL." I turned to the contents page and read chapter titles like "Buddhist Economics," "A Question of Size," and "Technology with a Human Face." I knew it was time to revisit these ideas.

*Small Is Beautiful* turns 50 in 2023, and the book is more relevant now than when it was written. Schumacher argues that all modern-era crises—poverty and inequality, environmental degradation, social isolation, to name a few—share a common cause: the "idolatry of gigantism." Humans are doing too many things too quickly on too large of a scale, and it's not sustainable.

In 1973, the United States faced its first major energy crisis since wealthy nations began using fossil fuels in earnest in the 1800s. In October that year, the Organization of Arab Petroleum Exporting Countries (OAPEC), led by Saudi Arabia, proclaimed an oil embargo. By March, the price of oil had risen 300 percent. Cars lined up at gas stations. The public was finally beginning to ponder the fact that fossil fuels, a finite resource, would not last forever, at least at the rates they were being consumed.

The country was also awakening to pollution crises. Los Angeles was covered in smog. The problem of radioactive nuclear waste was constantly in the news. Rachel Carson's *Silent Spring* shed light on poisonous chemicals like DDT, widely used at the time on farms. Might we wake up one year to a spring without birds because we'd poisoned them all?

Fast forward to today, and these problems have only compounded. Fuel prices remain volatile, and fossil fuel extraction and distribution are as politically fraught, on a global scale, as ever. Our rate of fuel consumption has doubled since 1980.[2] In addition to smog and radioactive waste, we now face climate change, water shortages in the West, contaminated drinking water, nanoparticle pollution . . . the list could go on and on.

Each of these problems requires big thinking and big solutions. But what if we thought smaller?

Schumacher presciently wrote that problems caused by gigantism won't be solved by gigantic solutions alone: "The economics of giantism and automation is a leftover of 19th-century conditions and 19th-century

View of Clay Bottom Farm. A productive farm need not be huge. With appropriate technology and a minimalist mindset, it is possible to earn a comfortable living on a small patch of land. Photo courtesy of Adam Derstine.

**Table 0.1.** Two Ways to Farm*

|  | *Small Is Beautiful* Farm | *Never Enough* Farm |
|---|---|---|
| **Leverage constraint** | Set specific limits to growth. | Perpetual growth is the goal. |
| **Farm with** *just enough* | Farm within ecological limits. | Extract value from nature and turn it into wealth. |
| **Essentialize** | Do less but better. Be selective. | Do more and sell more. |
| **Simplify** | Seek appropriate technology—tools with a human face. | Rely on complex, large-scale machines. |
|  | Integrate traditional wisdom for growing food. | Prioritize technological solutions. |
| **Localize** | Integrate local people for mutual gain between farm and community. | Maximize profits no matter the cost to community life. |
|  | Create jobs with human dignity. | Eliminate manual work through mechanization, whenever possible. |
| **End results** | Health, permanence, and beauty. | Sickness, degradation, and blight. |

* Ideas adapted from *Small Is Beautiful* by E. F. Schumacher

thinking and it is totally incapable of solving any of the real problems of today."[3] These problems will only be solved for good when we change our mindsets and lifestyles and policies to prioritize *less*: less extraction, less consumption; smaller houses, smaller cities, and smaller and more distributed farms. Schumacher points to Gandhi's "spirit of Swadeshi" as a model. If we organized ourselves into smaller economic units and lived more lightly on the Earth, we'd have a healthier planet, healthier people, economics without exploitation—and beauty.

As Schumacher stated, "Wisdom demands a new orientation of science and technology towards the organic, the gentle, the non-violent, the elegant, and beautiful."[4] Of course, Indigenous groups and many others in the United States have long pointed to another way of living. Schumacher urged seeking out traditional wisdom to pursue a right livelihood and to restore ecological balance.

While *Small Is Beautiful* is a book of macroeconomics, it is also full of practical advice. Schumacher's ideas undergird this book. In the pages ahead, I will show how we applied his concepts to our farm and tested them out in our community.

# A Small Is Beautiful Food System

My recent work with small farms and farmers from all over the world left me wishing that in the United States we had more—a lot more—small farms. Fewer than 1.3 percent of people in this country farm for a living, down from more than 50 percent in 1960.

To be sure, farming isn't for everyone. We need teachers and doctors and others who are devoted to specialty work. But we can do better than 1.3 percent.

Most countries around the world have us beat. In 2020 and 2021, I joined a USAID-funded project, implemented by Winrock International, in Nigeria. (I participated via Zoom because of the pandemic.) Nigeria has roughly the same population as the United States, however 7 in 10 people earn at least part of their income from farming.

In neighboring Chad, the rate is even higher at 80 percent. A Chadian exchange student, Christian Berambaye, who worked at Clay Bottom Farm in 2022, told me that almost everyone in Chad grows or raises something, no matter where they live. In his family's compound, they raise chickens and goats and grow mango, lemon, orange, and guava trees. Upon returning home, he plans to add tomatoes to the mix. Around the world, one in three people earn a portion of their income from farming. If you farm, there are probably neighbors and family and friends close by doing similar work. For my hometown of Goshen, Indiana, a midsize city of 40,000, to compete with a similar-size city in Chad, in terms of the number of farmers, we would need thousands more producers—up from approximately 10!

Cities around the world teem with small market gardens, roadside nurseries, and catfish growing in tanks on vacant lots, whereas in the United States, urban farming has been all but zoned out of existence. We had to apply for six zoning variances and plead our case to city officials before being allowed to farm in Goshen. More on this story in chapter 9, in the section "How We Dealt with City Officials."

While US farmers are few in number, the size of a US farm is comparatively gigantic. In 1850, an average farm sat on 80 acres. But now, the average farm comprises 450 acres, almost six times larger, and that number keeps growing. According to the USDA, farm size now doubles every 20 years. All the while, we keep losing farms—more than 100,000 in the last decade alone. In agriculture, the rule remains: *Get big or get out*, as former ag secretary Earl Butz infamously put it. In the 1970s, Butz sought to reform agriculture, to make it more friendly to agribusiness corporations. He wanted farmers to scale up, and he urged them to plant from "fencerow to fencerow." Sadly, his reforms largely succeeded—gigantism is still the trend.

To be sure, we could sit back and let a few enormous farms feed us—this is the path we are on. Instead, I think it's time to rapidly pivot the other direction, toward small and distributed farms as the foundation of our food system—otherwise, we miss out on what small farms have to offer. Consider this partial list:

- *Combatting climate change.* Small farms are more resource-efficient, in large part because small farmers ship food shorter distances. Up to 11 percent of greenhouse gas emissions can be traced back to big farms, an embarrassing statistic for Big Ag.[5] The best way to reduce pollution in the ag sector is to put small farmers in charge.
- *Less food waste.* Large farms routinely overproduce to ensure they grow enough to fulfill large contracts. In one study of California farms from 2019, 33.7 percent of marketable food grown was never picked.[6] Smaller farms can more tightly align their production with the marketplace, resulting in less waste.
- *Lower start-up costs.* Smaller farms require less land and equipment and fewer buildings, so lower- and middle-income folks can enter the profession more easily. The best way to increase equity in agriculture is to boost small farm access.
- *Stable food supply in times of disruption.* During pandemics, hurricanes, and war, a diversified small-farm-based food system spreads risk. During the early weeks of the COVID-19 pandemic in the United

Clay Bottom Farm heritage tomatoes.

12

States, corporate-scale farmers plowed in acres of onions and cabbage and other foods when their buyers abruptly terminated contracts.[7] Grocery store shelves emptied when communities panicked and rushed their local markets.

Consider an alternative, based on distributed small farms: In Ukraine, 8.3 million households—more than half the population—are growing 98 percent of the country's potatoes and 85 percent of other fruits and vegetables during a time of extreme disruption caused by the Russian invasion of 2022.[8] I discuss the unique plight of small farmers in the region in more depth in chapter 5, in the profile of a Romanian plum farmer (see page 103).

- *More biodiversity.* Small farms grow a wider variety of crops than gigantic farms; they preserve heritage and heirloom varieties passed over by the conventional food system.
- *More cultural diversity.* All cultures around the world, like the Mennonite culture I grew up in, have traditions rooted in food and farming. Small farms keep those traditions alive at a time when cultural distinctiveness, in many places, is under threat.
- *Populating food deserts.* Fresh healthy food, in too many neighborhoods, is not even available. This problem is sometimes called *food apartheid*, because humans, not nature, create these deserts. Micro and urban farms, located on tiny plots, supply healthy food where it's needed.

Plus, consider the fact that the food of the macro-ag world is crap. Much of it belongs in the garbage. It's giving us diabetes, heart disease, anxiety, and depression. *Real* food is grown on small diversified farms, not on giant monocrop operations. Many will argue that corn and soybeans and other such commodities are required to feed animals, but studies show that our animals would be better off eating grass and grubs and acorns, not industrial grains. We give them corn and soybeans because those are the cheapest, most convenient things to feed them, even though the result is nutritionally inferior meat and dairy and eggs.[9]

But can tiny farms really feed the world? The current global population of more than 8 billion is expected to rise to 9.8 billion by 2050. That's a lot of mouths to feed.

The truth is that small farmers are already feeding the world, especially in the global south, where population is growing fastest. As Vandana Shiva writes in *Who Really Feeds the World*, "While using only 30 percent of the world's resources, small-scale farmers provide 70 percent of the planet's food. Small-scale farmers, farming families, and gardeners feed us."[10] In the United States, our food system is dominated by a powerful few; outside of our country, small farmers in many parts of the world play the central role in feeding the people.

# The Layout of This Book

In part 1 of this book, I will explain the five principles we used to get small. Part 1 is not just for growers—it shows how anyone, on any type of farm, can do "less but better," to borrow a phrase from German designer Dieter Rams. At the end of each chapter, I tell the stories of profitable microscale farmers who exemplify get-small principles in interesting ways. For these case studies, I did not comb the farming world for idealized farms. Rather, these are ordinary farmers I happened upon who are doing an extraordinarily good job of leveraging the power of small. They are from all over the world, and the foods they produce range from catfish to tomatoes to plums. Put together, they are proof that it is possible to run efficient—and beautiful—farms on a human scale.

In part 2, I will show in greater detail how we put the principles into practice on our own micro farm, illustrating how we set ourselves up for high-flow production. I'll explain how to implement a deep-mulch system to simplify work and how to prepare growing beds in just two steps. I'll describe the seven field tools that we use, and I'll discuss how to electrify your farm and build better micro-farm infrastructure. You can read the chapters in part 2 as a quick-start guide for setting up a tiny minimalist market-garden farm. This setup, combined with a lean mindset, allows us to sell more than $85,000 per season from our ⅓ acre in production. The final chapter of the book shows an accessible plan for earning $20,000 from a backyard garden by growing four crops. My goal is to demonstrate how anyone on a tiny patch of land—as small as a yard—can run a successful farm.

I hope to inspire home gardeners to expand within the limits of normal-size urban lots to create spaces of food production. The ideas in this book can be used to manage a large garden more efficiently, perhaps allowing it to provide part-time income.

This book, then, is a blend of theory and practice, with an emphasis on practical solutions to food production on a microscale. As Schumacher notes, "An ounce of practice is generally worth more than a ton of theory."[11] I hope that these pages inspire millions of new producers to start up tiny farms on mini plots and that soon these farms will be found in every block of every urban residential neighborhood. This book is not just for farmers who already have a farm. It's for the farmers of tomorrow, who will plant seeds in small gardens and will feed the masses for a long, long time to come.

Five principles we use to earn a comfortable living from a tiny patch of land.

PART ONE

# Principles to Get Small

# CHAPTER 1

# Leverage Constraint

*Where is the rich society that says,* Halt, we have enough?

—E. F. Schumacher

The first step to farming with a get-small mindset is to determine three or four specific constraints, based on your values, to steer your farm. Once you know your constraints, write them down and leverage their power to do better, more focused work. We consider our constraints, which I share below, to be the key drivers behind our farm's efficiency and success.

But we didn't always feel a need for boundaries.

By the summer of 2013—our seventh year in operation—our farm was humming. We'd just finished building our fourth greenhouse and remodeling our processing room. We were farming full tilt, growing more than 60 specialty crop varieties—heirloom tomatoes, ginger, turmeric, figs, microgreens, pea shoots, yellow watermelon, and more—and selling to 50 CSA customers, 10 restaurants, and 2 grocery stores. We worked all week, including at a Saturday-morning and Tuesday-evening farmers market. We sold everything we could grow and worked as much as we could.

The farm was also a social hub. Interns and workers lived in a remodeled room in our barn and in our house. We hosted gourmet meals in our barn loft, partnering with a local chef. Customers for the meals came from as far away as Chicago. Each plate for these meals was prepared just in time. As the chef braised pork belly on a grill, I'd run out and harvest scallions and herbs to flavor the dish; minutes before the salad course, I'd clip pea shoots from the greenhouse. We also hosted large harvest parties for friends and customers. Our lives were rich and full and getting richer and fuller—and busier—every season.

Then, in 2014, Rachel gave birth to Arlo, followed by Leander a year and a half later. While we'd enjoyed our full farm life, we now needed to make room for kids, to carve out space to be together as a family, without being overwhelmed.

It was time to set some limits.

# Buddhist Economics:
# Right Livelihood on the Farm

In the 1940s and early 1950s, E. F. Schumacher was working as an economist for the British government. In 1955, he was given an economic development post in Burma.

In Burma, he saw proud, happy, beautiful people who lived simply even though they might have been poor by Western standards. He wrote to his wife, Muschi, that he wasn't sure he had anything to offer to the Burmese: "How can we help them, when they are much happier and nicer than we are ourselves?"[1] They were more generous and thankful. They seemed to go about their work in a relaxed way. They were clearly motivated by a different ethos than people in the West.

He began to study Buddhist economics and the concept of *right livelihood*, a teaching that undergirds Burmese life. In his first sermon after enlightenment, the Buddha explained that the way to peace and enlightenment is the Eightfold Path. The fifth "path," or requirement, is *right livelihood*—the idea that work is not solely a way to provide for your physical needs; work is also a spiritual practice.

As Buddhist teacher and author Thich Nhat Hanh explained it, "The way we earn our living can be a source of peace, joy, and reconciliation, or it can cause a lot of suffering." The goal of right livelihood is to find work "that is beneficial to humans, animals, plants, and the Earth, or at least minimally harmful."[2] Right livelihood says that there are limits we should place on our work in order to make sure our work is a source of social good and to preserve space for rest, time with family and friends, and time for social action and volunteer work.

In Burma, Schumacher saw right livelihood in action. In general, the Burmese consumed and worked less. Sometimes this was because of limited options or because it was the custom. But many Burmese people intentionally placed boundaries around their work. In Western cultures, Schumacher noted,

Spring radishes.

people work long hours under strain from economic pressures; in Burma, "you find that people have an enormous amount of leisure really to enjoy themselves."[3] The Burmese weren't antagonistic to physical well-being: "It is not wealth that stands in the way of liberation but the attachment to wealth; not the enjoyment of pleasurable things but the craving for them."[4] With boundaries on their work, the Burmese freed up time for creativity, for relaxing, for attending festivals and ceremonies—for living a right livelihood.

## Blending Customer Value with Farmer Values

We began our farm's transformation with questions. What might right livelihood look like for us? How might we intentionally create space for family time, fun, and non-farming pursuits?

For years, we had used lean principles to precisely identify what our customers valued. We visited chefs and asked them what they wanted from us, when they wanted it delivered, in what amounts. We took answers to these questions seriously and designed our farm around our customers.

Satisfied customers are foundational to an economically successful farm. If you have outside funding, you can pay less attention to customers,

Our boys, setting in lettuce transplants.

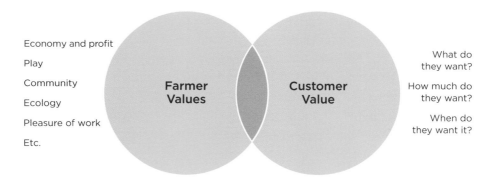

Economy and profit

Play

Community

Ecology

Pleasure of work

Etc.

**Farmer Values**

**Customer Value**

What do
they want?

How much do
they want?

When do
they want it?

but if not, the customers' needs are imperative to understand. You want to make sure you are producing goods that customers are willing to pay for.

On an intentionally tiny farm, you need to incorporate another set of values—your own. What kind of farm do *you* want? How big do *you* want it to be? What kind of farm appeals to *you*? How much time do you want to devote to the farm? Answers to these questions identify what you as a farmer value. They guide you to a set of boundaries.

Answers to these questions are not straightforward. Farmer values involve answering economic questions like how much profit you might need, but also how do you value play, community involvement, service, farming with ecological methods, and so on. Farmer-value questions are nuanced and can be difficult to answer.

But this work is essential. The core challenge of high-profit micro farming is to arrive at precise definitions of customer value and your own values, and then to farm where the two overlap.

## SETTING CLEAR BOUNDARIES

One afternoon, as our boys took an unusually long nap, we reassessed our business. We taped a giant piece of paper to a wall in our house and drew circles representing core tasks—like selling at the farmers markets, CSA distribution, bookkeeping, building maintenance, and so on. For each circle we discussed questions like how much time do each of us want to spend here? If the task generates income, how much is enough? What might "less but better" look like in our winter greenhouses or at the farmers market? Where should we cut back?

We had long talks about what to do. Admittedly, these were sometimes heated conversations. We were not always on the same page. Typically, I led the charge of trying to do more—"we should add another restaurant account"; "we should try these new varieties"—or at least keeping with our current pace, while Rachel held fast advocating for shrewd limits.

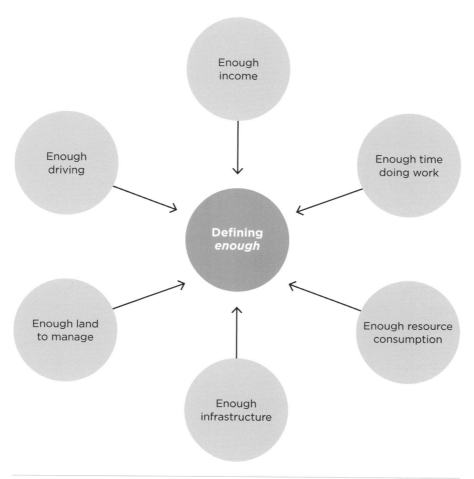

To self-impose constraints is really to define *enough*—a task at odds with our culture of "never enough." Defining *enough*—setting limits—does not need to set back an enterprise. It can actually propel better, more focused work.

In the end, I'm glad that Rachel's point of view held sway. She deserves credit. She pushed us to put less land into production, to keep focusing our business on higher-value crops, and to minimize our crop losses. We have moved our farm twice, and both times it was due to Rachel's persistence, wanting to make sure we were in the right location for what we wanted to be doing.

We continued the conversation over the course of a year, and eventually we created clear boundaries based on these talks. These boundaries helped us sort out our roles. I do most of the fieldwork; Rachel does almost all of our bookkeeping, and she sells our produce at the farmers market. While we each give advice to the other, the boundaries help us sort out who is primarily responsible, who makes final decisions.

The boundaries described in the next section, of course, are specific to our situation—to our preferences, to the types of crops we grow, to our location, and to our customers. There are certainly other ways a farm can set boundaries. I encourage you to determine what limits might be useful for you to define, in your own context. When you can, be specific. Precise boundaries are powerful tools for staying small. General targets are less motivating.

## THREE CLAY BOTTOM FARM BOUNDARIES

*1. To complete work in 35 hours per week or less.* We decided that 35 hours per week during the peak season of April through October would allow me time to manage the farm well while preserving energy for childcare. For Rachel, 10 hours per week, spent in bookkeeping and selling at the farmers market, would allow her primary focus to be on the kids while they are young. Our goal was to set a schedule where we could mutually support each other and create positive time to be together as a family.

Before kids, our work schedules were flexible. We would sometimes work late into the night to fill an order or catch up on the books. Now, we had defined boundaries. A key way that we reduced our own hours was to hire good help, with a flex-time system, which I discuss more in chapter 5 (see page 97).

*2. To grow on ⅓ acre or less.* We were farming on about 1 acre of land at the time Arlo and Leander were born. But if we were committed to cutting back our hours, we would need to reduce the acreage we were managing. In spite of our experience in growing, we still saw waste in the field. We knew we could do better. To reduce our size, we cut back on the production of lower-value crops and we doubled down on consistency in our methods to eliminate or greatly reduce crop failures. We became fastidious in our seed ordering, choosing strong and productive varieties. I set a goal of never having an open bed for more than two weeks, which meant that the growing area we needed could be compressed. I explain these and other changes we made to our growing system in part 2.

*3. To sell only in Goshen,* the small city where we now live. Before, we sold to accounts up to two hours away. This change allowed us to consolidate driving routes and save time on the road, reducing our carbon footprint. Less road time also means we save money on gasoline and vehicle maintenance. There is almost no downside to driving less.

Rachel inspecting tomato transplants.

This list is not exhaustive. We also set boundaries on the amount of plastic packaging and number of gas-powered tools we use, for example. As stated, keep the boundaries clear and specific, and write them down.

## THE POWER OF CONSTRAINT

To be honest, these boundaries sometimes made me nervous. They felt restrictive. I worried that less land in production—cutting it by more than half—would decrease our income. It didn't: We sell as much as before.

I worried that cutting back our time would mean we'd accomplish less. It didn't: We found creative ways to get work done on schedule, occasionally with time to spare.

I worried that selling to only nearby accounts would mean we wouldn't sell as much. That didn't pan out either.

On and on, every worry I had was wrong. I underestimated the power of a carefully set limit, adhered to with creativity and discipline, to inspire better work. On the surface, it would seem that a constraint, whether external or internal, is a negative force, restricting your range of options, holding you back, standing in the way of progress. That's a misconception. Clear limits help you do better.

Take tomatoes. When we planted too many, aphids overtook them because we couldn't keep up with a spray schedule. We got behind with pruning, and the plants grew wild, decreasing their productivity and slowing down our harvest. But with 30 percent less, we were able to keep up with pest control, pruning, cultivating, and so on. The patch was healthier. We were able to fill the same orders but with much less waste and work.

The power of a constraint lies in its ability to motivate positive change—in our case, to motivate us to get small. With limits, we were able to refine our structure toward a specific goal. Without specific limits, it would be too easy to muddle along and gradually succumb to "more-ness"—to keep getting bigger, working longer hours, because our culture and economy incentivize us to do so.

Constraints are the guardrails that allow you to live in alignment with your deepest values. In a business, constraints are a kind of pressure, steering processes toward greater value for the customer without sacrificing your own values.

While Rachel and I defined the parameters above, we didn't do so in a vacuum. We sought input from our farm team, and, in turn, our team became motivated to help us farm inside the limits. Nicole, our assistant farm manager, organized meetings with our accounts—the chefs and grocery accounts inside of our delivery radius—to pin down what we could be doing better. She worked with our accounts on creative solutions to reduce plastic. Everyone who helped in the field gave me ideas to squeeze more food into smaller spaces and to reduce crop failures.

We decided to keep our giant sheet of paper where we worked out our boundaries. It is a visual reminder that keeps us in check. If we are tempted to till more ground, put up a building, or sell to another account, we pull out the paper and ask, would this change fit within our boundaries? Or, alternatively, I just ask Rachel how she feels about adding another project. Either way, it's helpful to have a check.

Clarifying our limits ultimately incentivized us to move. We decided that a new location, closer to customers and resources for kids, would make our farming both easier and more enjoyable. In fall 2017, we began our search for a city farm.

# A Musician-Farmer in Colorado

*"Without a doubt, limiting our farm has made it
more efficient and more fun."*

Gregory Alan Isakov is a Grammy-nominated singer-songwriter from Boulder County, Colorado, who also runs a productive market garden. He succeeds in both careers by *setting clear boundaries*. Here is his story.

Isakov's Jewish ancestors worked as potato farmers in Latvia and Lithuania for generations. During WWII, for their safety, they moved to South Africa, where Isakov was born in 1979. In 1986—the height of apartheid—his family moved again, to Philadelphia, where he lived in a small apartment with both grandmothers and his brothers.

Isakov's split passions, for plants and for music, are not new. He started touring at age 16 and studied horticulture at Naropa University, in Colorado, right out of high school.

After college, Isakov lived out of his car for three years. "It was great," he told me. "You can actually get by with very little." He traveled around, working on small farms and nurseries and playing music at night. "I think I really just wanted to camp for free. I'd play a bar or coffee shop and make enough money for the campsite," he said.

But while his music career was ascending, he missed farming. So in 2015, he bought Starling Farm, just outside of Boulder, Colorado, and started to farm part-time. In his first season, he grew three open-pollinated corn varieties for a small seed company in New Mexico. Then he added lettuce and started selling to chefs.

Isakov now farms with three other people who work part-time on the farm. Six others live on the farm, and two of them have kids who hang out while the crew works. Isakov farms full-time in the summer and tours in the winter. When I talked with him most recently, in fall 2022, he was closing up the farm for the season and getting ready for performances in Vienna, Berlin, Stockholm, Copenhagen, London, Glasgow, and more than 60 other cities that winter.

## SETTING BOUNDARIES AT STARLING FARM

"Growing food is the ultimate manifestation of a co-creative life," Isakov told me. "Creating life. Moving with life. Reacting, adjusting. It's the same with songwriting. I find them both to be balancing."

The key to the balancing act: good boundaries.

In the early years, Isakov and his team grew food for a CSA program, chefs, florists, and a seed company. "We had a hard time setting boundaries," he told me. They rarely had days off because their customers needed their food at different times. There was a relentless stream of orders to be filled. Burnout was on the horizon.

They decided to define limits. They stopped selling to restaurants and focused on their CSA. They simplified the CSA program so that they would offer only one size of box, with no customization.

Next, they decided, counterintuitively, to take off for two weeks in the middle of the season. "The biggest time for burnout, for us, is the first or second week of July, every single season. We're tired. It's hot. It's hard to get things to germinate. It's extra work for everything." They divided the CSA into a spring and fall season with their two-week vacation in the middle.

They moved CSA pickup off the farm, so they wouldn't have to host the traffic of CSA pickup. Limiting traffic streamlined work, and besides, "We're

Gregory Alan Isakov with a bundle of beets. Photo courtesy of Rebecca Caridad.

private here," he said. Finally, they eliminated crops such as rainbow carrots that were time-intensive to pick but sold in low volume.

## BENEFITS OF CONSTRAINT

The changes gave Isakov the benefits he was looking for. By constraining the farm, Isakov has energy for other pursuits. He said:

> *It's important for me to have another job, to take the pressure off. That lets me enjoy the learning process on the farm more. We never had days off in previous seasons like we do now. I love writing and I love art and I love growing and I love music. When I'm on the road in the wintertime, I get excited*

Overhead view of Starling Farm. Photo courtesy of Gregory Alan Isakov.

*about my seed order, and I sometimes order seeds on tour. When I'm here [on the farm] all season, and I can't leave because my plants will die, I get excited about going out.*

With strategic boundaries, he reaps the rewards of both careers:

*In farming and in music there's so much conversation about profit and loss, inputs and outputs, about economics and efficiency, which makes sense. But what's left out is the economics of health. Joy. A good night's sleep. Working outside with your friends. Time to hang out with your family. These should be as important as profit. It's a balancing act.*

# CHAPTER 2

# Build *Just Enough*

*For every activity there is a certain appropriate scale.*

—E. F. Schumacher

One morning at our farmers market, a regular customer stopped by our booth. He picked up three tomatoes, paid for them, and put them in his bag. Then he looked up and said, "I've got an idea for you."

He asked if we wanted to buy land that he owned. "It'd be perfect for a farm," he said. He unknowingly hooked our interest when he added, "You'd be right in town."

So began our new farm search. We were ready, after two seasons of farming with kids, to transform our business—to right-size it for a family, on a smaller plot closer to town.

We visited the plot, on Hackett Road, three or four times, digging around to inspect soil, imagining buildings and gardens here and there. We meandered one morning to see where the sun would shine first, and we returned at dusk to see where trees cast their shadows. We inspected other properties, too, but we liked Hackett Road. It was inside the city—about 1 mile from downtown—with loamy soil, open access for utilities, and in a beautiful spot. We signed papers in the fall.

That winter, we excitedly sketched out our dream farm. This move was a chance to build a tidier, less-but-better farm. We met with an agronomist who helped us understand the soils, slopes, and water tables on the property and to pick the best spot for growing vegetables. We worked with an engineer to begin drafting plans from our napkin sketches. In this chapter, I'll explain the principles that guided us as we designed our new farm. In chapter 9, I discuss the spaces we built in detail.

A tiny-scale farm is most productive when it has just enough infrastructure: not too much, not too little. Photo courtesy of Adam Derstine.

# The Principle of Just Enough

Our primary goal was to build just enough—not too much nor too little. *Just enough* means we have the infrastructure required to produce efficiently and support ourselves economically, but no more than we absolutely need.

According to Schumacher, when designing new things, "There is an irreversible trend, dictated by modern technology, for units to become ever bigger," and we rarely ask, "What is actually *needed*."[1] When setting up a farm, it's easy to overdo it: to build excessively large buildings, till more land than you can manage, and buy tools you won't use.

But excess has a cost. Too much stuff and infrastructure actually decrease speed. Too many buildings, hydrants, and tractors slow production because your time is gobbled up in maintenance, in looking for the right tool, and in maneuvering around things.

On the other hand, it is possible to take minimalism too far. With too little, you can't produce efficiently. Crops go to waste, for example, without adequate refrigerator space. Without enough greenhouse room, you can't supply food during the critical shoulder seasons, when demand is highest. In our

## The Principle of Maximizing Fixed Costs

Once you have what you need, then maximize its use in order to make the most of your investment. In a lean system, this principle is called "maximizing fixed costs." Fixed costs are long-term investments in things like warehouses, sheds, tractors, and building materials; you pay for these things no matter your level of production. Variable costs, on the other hand, go up and down as your production fluctuates: think seeds, fertilizers, fuel, and packaging. Maximizing fixed costs means using investments at full capacity.

When I was a teenager, I worked with small farmers in Guatemala. Their buildings were in constant use. Simple sheds stored equipment, sheltered food for drying, and, if needed, were emptied out for community meetings. There was rarely spare sheet metal laying around—extras were put into use on a roof somewhere. Investments were maximally used. By contrast, on many farms in the Midwest, empty sheds and never-used building materials clutter the landscape.

Maximizing fixed costs requires a certain mindset. In 2020 to 2021, as stated, I worked with a USAID-funded project in

Nigeria, with catfish, soy, maize, rice, and cowpea farmers. Project designers there wanted to try a new approach. Before looking to solutions that might require more infrastructure and investment, they first looked for ways to reduce waste, to maximize the use of buildings and equipment and land that farmers already had.

Team members asked farmers, "Where do you see waste? When do crops rot in the field or in storage? Where is there wasted motion or overproduction waste? What ideas do *you* have to improve?" The farmers were quick to answer. They offered ideas for feeding fish more precisely, threshing rice more efficiently, cultivating maize with less motion, and so on—all ideas for better utilizing fixed costs on their farms. The story of Glory Adonase on page 44 is one concrete example of this mindset put into action.

While an instinct to do more with what you already have prevails in economically poor countries, there is no reason the ethic can't apply anywhere with the same benefits: less waste, higher profits, and fixed costs put to full use.

early years, we weren't efficient in large part because we grew our business faster than we could support it with infrastructure. On a few occasions, I remember frantically driving fresh-picked crops into town to a rented walk-in cooler because we didn't have enough space in the farm's cooler.

Between too much and too little is the sweet spot, the middle ground where economically viable, sustainable, and beautiful solutions are found. This is what you actually need.

# Just-Enough Principles from Preindustrial Farms

Many of the early sketches that Rachel and I drew for our new farm were wildly impractical, with too many buildings, a cluttered layout, and no cohesive aesthetic. We wanted to build our new micro farm according to a just-enough ethic, but what, specifically, did that mean? How should we orient the buildings in relation to one another? What building materials should we use? Does just enough mean 1½" or 2" underground water lines? Does it mean six hydrants or eight? Does it mean a greenhouse that's 24' or 30' wide? Should the processing room have one drain or two? Does just enough mean solar panels? A heater in the garage?

We needed more guidance. Schumacher encourages looking at traditional models as a starting point. So, we picked up old family photo albums to see the types of farms our ancestors built, and we read books on traditional farm design, to combine with our own experience. Before the industrial revolution, which accelerated in the 1800s, nearly all farms were small by today's standards. Few farmers had the means to "get big." One person could not farm thousands of acres because the technology for gigantic-scale farming did not exist. Hence, ideas for compact farm design abound if you look back far enough.

Here are farm designs that gave us more direction.

## THE TRADITIONAL GERMAN FARMHOUSE (*MITTELDEUTSCHES HAUS*)

Ancestors on both sides of our family are Mennonites from Dutch- and German-speaking regions of central Europe. They fled their homelands in the 1500s because of religious persecution and eventually ended up in North America. Wherever they went, they built farms.

A traditional German farmhouse featured a large central mudroom that separated a working area, with animals and feed, from the bedrooms, living space, and kitchen. I remember my German-born grandmother telling me stories of milking goats—right in the house! The pail she used sits in my office.

This design inspired us to combine work and living spaces into one barn-house. This would save material and labor cost in the building process and heating and cooling costs down the road. Well-designed multifunctional buildings can create smoother workflow than when rooms are separated. We decided to put our processing room, propagation house, spray station, and residence into one building. The ability to go from house kitchen to processing room to propagation house under one roof is magic.

Around 1790, Mennonite relatives on Rachel's side of the family settled in the steppes region of the Russian empire (now Ukraine) before being forced to move again during the Russian Revolution in the early 1900s, settling eventually in Idaho. We still eat borscht (beet soup), *zwieback* (Russian rolls), and *pluma moos* (cold plum soup) during the holidays, traditions passed forward to us from her parents and grandparents.

The Molotschna colony, where Rachel's ancestors lived, featured 57 highly organized Mennonite villages. Family homesteads were built close to one another, and homes contained spaces for multigenerational living

A farmstead in the village of Blumenort, Molotschna, possibly taken between 1901 and 1920. The barnhouse building consolidates many functions under one roof. Note the simple architectural lines. The owners are standing in front of their home. Photo courtesy of 44-261 Gerhard Weiss farmstead, Gerhard Lohrenz photo collection, Mennonite Heritage Archives.

and for hosting visitors. This design inspired us to add a small loft apartment for family and visitors. (Some Mennonites remain in Molotschna, where Russian forces began occupying territory in 2022.[2])

The Ira Harvey Homestead, with connected house and barn, in Warner, New Hampshire, is now a bookstore. It was built in 1795.

## THE CONNECTED FARMS OF NEW ENGLAND

Some of the first homesteads built by European settlers in the northeastern United States were "connected farms," consisting of many farm buildings attached to one another.[3] The design allowed farmers to feed animals during cold northern winters without having to step outside. The farm buildings, often oriented east to west, created a microclimate on the south side for a kitchen garden and barnyard. These farms inspired us to use the placement of our buildings to block wind and create sun traps for our crops.

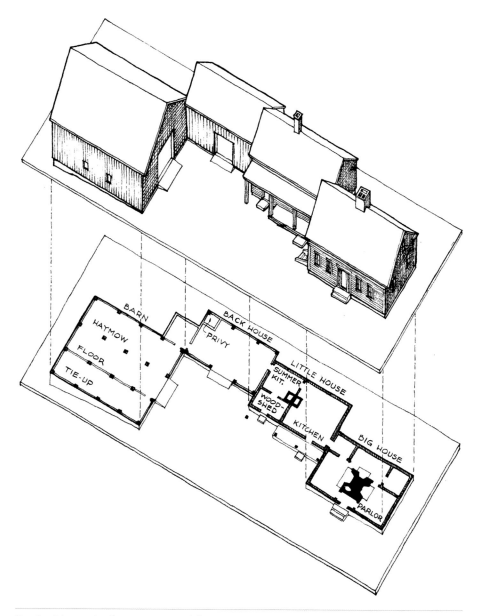

Typical layout of a connected farmstead, where farm buildings are used to block cold winter winds. Illustration by Thomas C. Hubka from *Big House, Little House, Back House, Barn* (Brandeis University Press, 2022).

They also inspired us to aim for balance and order in our buildings—to incorporate aesthetics. The connected farms appear balanced, even though rooflines and building shapes vary. They are interesting and beautiful. According to historian Thomas C. Hubka, architects of these early farms purposely sought to combine function and aesthetics. He writes:

*The idea that function and beauty (or order) were inextricably united is one of the most fundamental tenants of a folk or vernacular value system. The ideal of beauty in the folk system is intrinsic to the work of everyday life and is never detached from that life. The beauty of the farm flower is, therefore, integrated into the productive kitchen garden. The ornamental tree may be beautiful, but it also bears edible fruit and provides a border for productive fields.[4]*

## THE EDO-PERIOD JAPANESE FARMHOUSE

The Edo period in Japan lasted from 1603 through 1867 and is looked to as a model for what sustainable design might look like in the future. During those years, Japan was isolated from the rest of the world, and self-sufficiency was the key feature of rural life. Households grew their own rice and vegetables, raised their own meat, and made their own clothes, pottery, furniture, and tools. Their farmhouses, like the traditional German *mitteldeutsches Haus*, featured multifunctional spaces that integrated work and home life. In fact, many traditional Japanese houses included a *doma*, a workspace integrated into the house, which sometimes took up half of the overall interior area.

Nearly everything on these farms was recycled or reused at the end of its life. Worn-out clothes, shoes, and even the top few inches of the dirt floors were often turned into compost, along with all human and animal manures. The Edo-period farmhouses inspired us to build sustainability into our design. We decided to power the farm with solar panels and to use reclaimed materials as much as practical.

# Dieter Rams Down on the Farm

Dieter Rams is a retired German designer most closely associated with Braun products and with the functionalist school of industrial design. He is not a farmer, but his ideas, I think, would work well on a farm.

A traditional Japanese farmhouse. Note the high pitch of the roof, which allows attic space to be fully utilized. Photo by Getty Images/nicolasahan.

## Just Enough Is Simple

Each of the preindustrial designs we took inspiration from featured simple floor plans and architecture, often a single roof-line. Surely part of the reason is that farmers themselves typically did the construction, often with limited resources. Simple designs economize time and building materials. This is not to say that spaces were not sometimes sophisticated—they often featured carefully designed porches, niches, and walk-through spaces that added functionality—but their overall designs were simple enough for owner-builders to construct.

Simple layouts appealed to us. We designed our barn-house with four outside corners and a steep-pitched roof, with a single peak, to shed the snow. The high pitch of the roof also allows enough room on the second story for living space, a design feature in each of the preindustrial models in the previous section.

Like Schumacher, Rams believes that people should be at the center of design. He said, "The indifference towards people and the reality in which they live is actually the one and only cardinal sin in design," and that "Design should not dominate things, should not dominate people. It should help people. That's its role."[5]

Rams popularized the phrase "less but better," and his overarching goal was to make products useful through their simplicity. His designs have a timeless quality, with simple lines and intuitive features. He is famously quoted as saying, "Quiet is better than loud." Here is Rams's list of good design principles; I've added a farm twist.

1. *Good design is innovative.* The best farms are imaginative—they incorporate new technologies for growing food and push the boundaries of existing farm models.
2. *Good design makes a product useful.* At its core, a farm should be useful. It should be designed so that fruits, herbs, flowers, and vegetables flow smoothly from field to processing room to delivery van.
3. *Good design is aesthetic.* The physical environments we inhabit affect our well-being, thus a farm should be a pleasing place to live.
4. *Good design makes a product understandable.* A farm should be as self-explanatory to use as possible. Propagation houses should be located near growing beds, delivery vehicles parked near coolers, and tools hung at eye level near their points of use.

5. *Good design is unobtrusive.* A farm should blend as seamlessly as possible with the environment around it. It shouldn't try to dominate the landscape.

6. *Good design is honest.* A farm should have nothing to hide from visitors. In fact, it should be designed with public walk-throughs in mind.

7. *Good design is long-lasting.* Good farm design should be timeless. Durable materials should be used as much as possible. It should be planned for generations of farmers to use, not for obsolescence.

8. *Good design is thorough down to the last detail.* A farm should not be slapped together. Care and accuracy in the farm's design shows respect to the plants, animals, and people who inhabit it.

9. *Good design is environmentally friendly.* Farms should be at the cutting edge of climate remediation, carefully designed to sequester more greenhouse gases than they generate.

10. *Good design is as little design as possible.* Well-designed farms should be as simple as possible, with buildings that are easy to construct and repair.

## Essential Spaces at Clay Bottom Farm

After a season of planning, it was time to put the ideas in this chapter into physical form. Below, in broad strokes, is how we synthesized these principles and built a farm. In chapter 9, I'll explain in detail how you can set up infrastructure like this on your own micro farm.

We started our farm build by salvaging quality building materials, inspired by the sustainability of Edo-period farmhouses. In September 2017, Josh Janzi, a Clay Bottom Farm worker at the time, and I demolished an old corn crib—a tall building with wooden slats for air-drying corn—that sat unused at the farm where I grew up. So, we began building our new farm by taking down part of an old one.

We pulled the crib down with a tractor and chains and used an air-powered denailing gun, borrowed from a friend, to remove nails. We stacked the wood on hay wagons and drove them to the Hackett Road property. The corn crib was originally built with lumber that my grandfather sawed from the surrounding forests. It was rough cut, and most pieces were twisted or warped or both. But it was also hardwood and thus durable and beautiful. We eventually used the wood to build porches and for trim in the barn-house.

Our build started with an act of destruction. To salvage lumber, we tore down a corn crib from the farm on which I grew up.

## AN INNOVATIVE GREENHOUSE

Next, Josh and I started work on the greenhouse. Inspired by compact traditional farm design, Rachel and I decided to place the greenhouse as close as practical to our barn-house, about 12 steps away.

Following Rams, we wanted the greenhouse to be both simple to use and durable. We installed deeper posts for stability, sidewalks on both ends for tidiness, bifold doors set into the end walls, and extra-tall sides for ventilation and so that we could enter from the sides. We added a peak ventilation system at the top to let hot air escape.

Josh and I spent the winter building this greenhouse. We worked with bare hands because the system was held together with nuts and bolts and other cold metal parts, which we could only grab with fingertips. We carried thermoses of hot water with us to soak our hands every few minutes.

In March, we pulled the plastic over the arches. We assembled another crew of friends to help us pin down the plastic in one afternoon. The next morning, we seeded our first crops—greens and spinach—into the ground. It was an exciting moment after a long winter.

## A MULTIFUNCTIONAL BARN-HOUSE

In January, our barn-house foundation was finally poured, and an Amish crew framed and sheathed it.

We were inspired by traditional German farmhouses to compress many functions into this one building. Specifically, we placed a vegetable processing room, the most-used space on the farm, on the west side so that the

We built our first greenhouse in the winter, warming our hands in thermoses of hot water. The 34′ × 148′ structure features a peak vent and 6′ tall sides. More than half of our crops are grown under protection.

space would stay cool in the morning when we wash vegetables. It shares a mini-split electrical heating and cooling system, a water heater, and drain lines with the rest of the building, saving costs. We attached a propagation greenhouse to the south side and a lean-to for hosing off vegetables and

The barn-house during and after construction. Inspirations from preindustrial farmhouse designs include a high-pitched roof, simple lines, and a consolidation of many functions into one building. A propagation greenhouse is attached on the south side.

loading our van just off the processing room. Our living quarters are on the east half of the building and a farmworker/guest apartment is on the second floor. This combines work and living spaces, like Edo-period and early Mennonite farmhouses.

A cement-floored mudroom, which is also a workroom where we store seeds and supplies, divides the barn-house in half, creating clear dividing lines between living quarters and business. We created an alcove with a pedestal for flowers next to the entry door to the living quarters so visitors to our home know where to enter.

Following Rams, we valued *honest* design. We made a rule that as much as possible we wouldn't use materials that hid what they were: no plastics trying to look like wood or laminates trying to look like stone. We clad the house with sheet steel siding.

On our porches, we used posts and beams and old roofing metal from the corn crib, with no sanding or painting. We trimmed the doors and windows inside the barn-house with corn crib slats. We installed floors in most of the house using low-cost red pine boards from a Michigan sawmill. The pine is full of knots and imperfections, which please us but sometimes confuse others. One contractor, thinking that our pine floors were the subfloor, asked us when we planned to install the "finish flooring," which made us laugh.

Through the summer of 2018, we farmed part-time while we built this barn-house. Building was a community effort. My dad and I finished the drywall, and our siblings helped us paint. An Amish teenager and I installed the pine floors and the trim. Local Amish farmers helped us clad the house.

A window connects this propagation greenhouse to the rest of our house, allowing excess heat to warm the barn-house in the winter.

A local Amish cabinet shop turned corn crib floor boards, beaten up as they were, into beautiful kitchen cabinets.

We didn't get everything right in our build. The greenhouse turned out to be not quite large enough to supply customers with greens all winter and tomatoes all summer. We've since added another smaller one. At almost every turn, we were tempted to build more than we needed, and in retrospect, we could have saved space and built smaller. We were warned by friends that building a house is a learning experience. When you finish, they told us, you will be tempted to start over and do it again. They were right! While we don't plan to start over, we do plan to keep improving our farm.

Every property is unique, and there is no one right layout for every farm. But for tiny farming, start with just enough—not too much and not too little—and lean heavily on design inspiration from preindustrial times. Learn from these earlier farmers: Stay small, then maximize what you build.

# A Catfish Farmer in Nigeria

"I feel like I am following in the footsteps
of my parents and grandparents."

Glory Adonase carefully designed a high-flow catfish farm with a just-enough mindset. Her infrastructure is compact and fully utilized. Some of her catfish tanks are even in her house. Here is one approach to catfish farming she wanted to share that demonstrates her mindset.

## ROOTING OUT WASTE IN CATFISH PRODUCTION

Throughout summer 2020, Nigerian extension agents arranged meetings with catfish farmers in two critical catfish-growing states in Nigeria—Delta and Cross River. They asked the farmers for ideas to improve production by removing waste.

The interviewers were members of the USAID/Nigeria Feed the Future Agricultural Extension and Advisory Services Activity, and the initiative was funded by USAID and implemented by Winrock International.

Catfish farming is expanding quickly in Nigeria; the Niger Delta (including Delta and Cross River states) provides more than 82 percent of the country's fish production.

You don't need to own much land to raise catfish. Many catfish farmers live in cities and grow catfish behind their homes or on vacant lots. But it is not cheap to start. Catfish ponds can cost more than $1,000 (N460,000) and concrete tanks as much as $3,700 (N1,700,000). Farmers usually self-finance, perhaps saving for years, or use family gifts.

The farmers were eager to share ways to cut out waste and maximize their investments. They named one problem most frequently: high mortality of young fish.

## THE STORY OF GLORY ADONASE

Glory Adonase is the granddaughter of subsistence farmers who grew yams and vegetables. Her father was a seaman, her mother a teacher. "I feel like I am following in the footsteps of my parents and grandparents," she told me.

She entered university to study marine biology, but a professor encouraged her to switch to catfish farming instead because of the economic opportunity. One day, a fish breeder came to the university to give a lecture. Adonase walked up to the man after his talk and said, "I want to work for you." He said, "OK, but I can't pay you."

She went to work for the breeder for free, traveling to the farm every Saturday then delivering fingerlings—starter catfish—to other farms. When the breeder moved to another state, customers came to her asking for fingerlings. Thus, at age 25, she started her own business.

Adonase called her farm the Bangadonase Hatchery. Her business grew quickly, but she wasn't satisfied with her process. She told me that many of her fingerlings were dying in transit or soon after reaching customers, who would become unhappy with her.

"When a member of the Extension Activity team asked me if I'd like to participate in a project to reduce fish mortality, I quickly got into it," she told me.

Glory Adonase in her catfish hatchery. Photo courtesy of Extension Activity.

Extension agents introduced Adonase, along with a cohort of other hatchery operators, to a new method—raising young fish to a juvenile stage before selling them (fingerlings are typically three to four weeks old; juveniles are six to eight weeks old). Adonase implemented the techniques right away. Her mortality rate dropped from 30 percent to 3 percent.

Adonase's customers were thrilled. Because juveniles are older, the farmers could now operate three (and up to five) production cycles per year instead of two. Fish from juveniles grow more consistently, thus requiring less sorting. The new technique better utilizes everyone's fixed costs.

When I last spoke with her, Adonase said she'd added 20 new customers in the past three weeks alone and that she recently ran out of stock. She increased her sales volume last year from 20,000 juveniles per cycle to 120,000 due to a combination of lower mortality and better marketing. (She has started sending bulk SMS, WhatsApp messages, and MIP video messages to farmers in her network, following a training by the Extension Service.) Better process "has really given me an edge," she said.

To keep up with demand, she is partnering with other start-up hatcheries in the cohort that was set up by the Extension Service. "When I am out of stock, I sublet the supply," she told me. Her tanks have never cycled through more fish than they do now. She is at full capacity, right where she wants to be.

# CHAPTER 3

# Essentialize:
# The Pareto Principle

*Progress cannot be generated when we are
satisfied with existing situations.*

—TAIICHI OHNO

In 1906, the Italian economist Vilfredo Pareto found a pattern. He observed that 20 percent of the population of Italy owned 80 percent of the land. When he studied land ownership in dozens of other Western countries, he found the same 80/20 pattern. How could so many countries, he wondered, share such an eerily similar statistic? Pareto spent the rest of his career trying to figure out why that pattern existed.

In 1941, American management consultant Joseph M. Juran happened upon a book by Pareto. Juran specialized in quality control. In his own line of work, he saw the same pattern: 80 percent of the quality problems were the result of 20 percent of the causes.

Juran wrote about his and Pareto's observation. Suddenly, people from all kinds of professions started seeing the pattern, too. Hospital managers noticed that 80 percent of costs were incurred by 20 percent of patients. Retailers noticed that 80 percent of their sales were generated by about 20 percent of their customers. Computer programmers observed the pattern, too: 80 percent of bugs resulted from 20 percent of the code.

The core task of an essentialist farmer is to choose the right crops to grow and the right customers to sell to.

**The Pareto Principle**

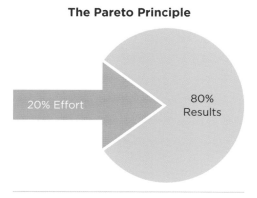

The 80/20 principle states that, in many endeavors, just 20 percent of efforts produce 80 percent of results. On many small, diversified farms, this principle holds true.

The 80/20 principle applies outside of the business world, too. Think about your email: 20 percent of most email inboxes contain valuable information while the rest is junk mail. On a recent vacation, we took stock of all the places we went and realized that 80 percent of the enjoyment on our trip probably came from about 20 percent of our activities. The 80/20 rule, in summary, says that in a preponderance of events, 80 percent of the outcomes are the result of 20 percent of the effort.

Importantly, this rule, popularly called the Pareto principle, is an *observation*, not a law of nature. There are exceptions, and in almost no case is the distribution of effort and reward exactly 80/20. The principle is simply a pattern that has been noticed and documented for many decades in a plethora of environments.

What the 80/20 observation points to is that there is always an imbalance of inputs and outputs. Some initiatives will always produce better results than others.

## Pareto on the Farm

In November 2018, we moved into our new barn-house. By then, we'd clarified the limits we wanted to place on our business, and we'd set up a just-enough physical environment to support our new vision. Now it was time to rethink *process*.

I was nervous. Could we really earn sufficient income on such a small footprint? Would this downsizing experiment work?

On this new farm, we'd cut our workspaces nearly in half compared with before. We'd decreased our greenhouse growing space from 9,000 square feet to 5,000. We'd shrunk our processing room from 1,000 square feet to 420 square feet. We'd eliminated our 8′ × 8′ walk-in cooler and replaced it with a three-door refrigerator. We planned to farm ⅓ acre compared with 1 acre before. In other words, we essentialized our physical environment. In this chapter, I'll explain how we essentialized our farm management.

On a small farm, the Pareto principle applies to products, to customers, and to activities in the field. My own observations line up with Pareto and Juran's: About 20 percent of the products generate 80 percent of the income, and 20 percent of the customers provide 80 percent of the cashflow. In the field, a lot of our work is truly necessary, but a lot could be reduced or simplified. I'll talk about how to apply 80/20 thinking to field practices in the next chapter. Here, let's look more closely at products and customers.

## WHAT ARE VITAL PRODUCTS?

To qualify as vital, a crop has three bars to meet:

1. You can produce it in high volume.
2. You can grow it at a low cost.
3. You can sell it at a high (fair-market) price.

All three factors are essential. You can't just focus on volume—growing as much of a crop as you can—while paying scant attention to costs and price. As a simple example, let's say I grow $50,000 worth of sweet corn, which represents half of my sales. It would seem that sweet corn is a vital crop. But to grow that much corn, I might have incurred $45,000 in costs (I admit I am exaggerating to make a point)—to buy seeds, to hire pickers, to buy packaging, and so on. Unless I can sell it for a higher price, the corn barely pays. Volume means nothing without low cost and high margin (your profits, after subtracting your costs).

On the other hand, low costs and high margins are meaningless without volume. If I sell just two or three heads of lettuce at a farmers market, the sales doesn't justify the hassle—to pick the crop, to make space for it at the booth, to create a sign—even if the margin was high.

We keep a home garden where we still grow smaller amounts of butternut squash and watermelon, sweet potatoes, green beans, edamame, and a whole lot more simply because we want to. We've planted apple, peach, pear, persimmon, and fig trees. We've put in a nut orchard with hazelnut,

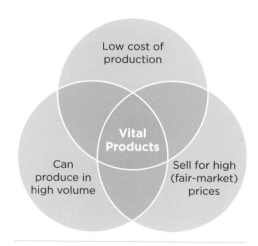

An essentialist farmer focuses on products that are low in cost to produce and that sell in high volume for high prices.

pecan, and chestnut trees. But we keep the line clear between production and home-use crops. With a profitable business, we have more time and energy to focus on the homesteading projects that we enjoy.

## WHO ARE VITAL CUSTOMERS?

In a preponderance of businesses, a small number of customers (perhaps 20 percent) account for most of the income. Those are the vital customers to focus on. A vital customer buys consistently; pays a fair, living-wage price; orders in relatively large volumes; and is low cost to deliver to. Choose them wisely, and you will be in business for a long time.

A vital customer is not simply one that pays you the most over the course of a year. Costs also must be considered. If you sell online, there is a website fee. At a farmers market, there is probably a booth fee and of course your time. With CSA and restaurant sales, there is the cost of communicating with customers and chefs and the cost of delivering your food. A vital customer is one who orders in a volume high enough to justify the costs required to sell to them.

Also, consider the hassle. Do you sell to a chef who puts in large orders but only pays after weeks of wrangling? Do you have a restaurant who will only take deliveries outside of your typical work hours? These are factors to consider, too.

Finally, consider prices. Just because customers order in volume doesn't mean they pay prices that offer you a fair compensation for your work. Even with shrewd business practices, margins on small farms can be razor-thin. Few local farmers can afford to sell large volumes at discount prices. It's better to expect fair prices.

You don't have to please everyone. Your business will take a huge step forward when you learn to discern which customers are vital for you and which ones are gobbling your time and resources.

This is not to say that there is no place for selling to a restaurant just because you want to or for selling at a discount to a charity cause. Your own values are still relevant. Maybe you want to support an upstart chef who

The most vital customers for tiny farms are those located closest to the farm who order in high volumes and pay fair prices.

Selling food at the Goshen Farmers Market.

## Overdevelop Your Markets, then Get Choosey

A smart practice with customers is to generate more sales leads than you actually need. For example, in 2020, we grew high-CBD hemp and turned it into tinctures. To find sales outlets, we called and visited dozens of shops that already sold hemp tinctures. Some were close to home and others were two hours away in Chicago. We got our product into the hands of several stores, but since then, we've edited. We now sell to just one store, located close to home, that pays us on a regular basis and orders in sufficient volume.

We do the same with restaurant customers. We've developed relationships with more than 20 restaurants over the course of the last 16 years. But we now sell to only a handful of carefully chosen accounts that are close to home, pay us on a regular basis, and put in consistent, large orders. If one of our vital restaurants goes out of business, we can always fall back on another account.

## Visualize Vital Crops and Customers

Visuals can help to identify your vital crops and customers. For example, create a bar graph where you line up crops from highest to lowest in sales revenue. The graph might surprise you. It might reveal that some crops drained your energy but only sold in small volumes.

Another useful visual is a bubble chart, where large bubbles represent crops that sell in high volume and small bubbles are low-sellers. Do the same with customers. Seeing your business graphically like this opens your eyes to the fact that equal efforts don't produce equal outcomes.

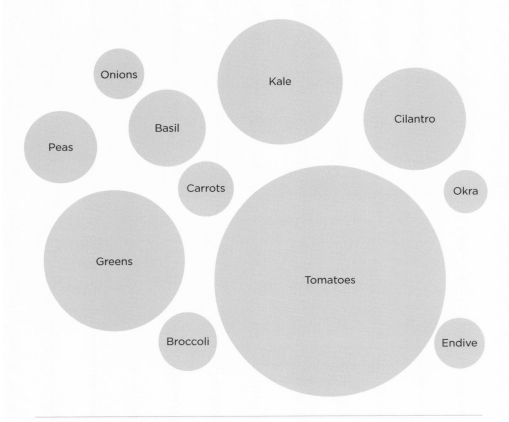

Bubble charts can reveal the distribution of effort and reward.

can't yet afford to put in large orders. Or perhaps you want to donate a portion of your food to a food pantry. We sell at our farmers market (though less frequently now than before) because we enjoy it as a social outlet and because, through programs like SNAP (Supplemental Nutrition Assistance

## Farming the Vital Core: Using Pareto to Get Small

It is worth repeating that you don't have to do it all. By choosing to do a smaller amount of things well, you can actually increase your revenues while doing less. The trick is to choose the right things to do.

If your goal is to grow as big as possible, then by all means go for 100 percent. Keep growing all of your revenue streams at the same time. Don't discriminate between them; just do more of everything.

But if you want to work less, focus on your *vital core*—the vital crops that overlap with vital customers. In all likelihood, unless you are operating at a large scale with hundreds of accounts and products, your vital core is probably surprisingly small.

To be sure, cutting out customers and products in the 80 percent category might seem cutthroat. But the best way to propel yourself forward while shrinking your workload is to ruthlessly focus on the crops and customers that carry a disproportionate weight.

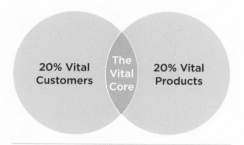

The vital core is the overlap between vital customers and vital products.

Program), we can offer our food at discounted prices to those who couldn't otherwise afford it. It's completely valid to structure your business around your values and not just your pocketbook.

## The Pareto Principle at Clay Bottom Farm

Before the first growing season on our new farm, Rachel and I listed our crops and our customers, and we started to make hard choices about where to focus. We eventually settled on the crops and customers that I list on the next page. This Pareto-inspired editing has paid off. We earn as much as before, though we sell to fewer accounts and work fewer hours.

### THE VITAL FEW CROPS AT CLAY BOTTOM FARM

Before our move, we were growing more than fifty kinds of vegetables. At our new farm, we chose the five focus crops—our vital few—with four secondary crops that are still important but not quite as essential to our bottom line. We used Quickbooks (our accounting software) to run sales reports, which helped us identify top sellers. We also considered yield per square foot (on our micro farm, we expect at least $5 per square foot) and the time each crop requires to harvest and pack (we aim to harvest and pack $150 worth of product per hour).

Here are our five focus crops:

1. Tomatoes: heritage, red, yellow, cherry
2. Salad mix: mostly lettuce in summer, mostly greens in winter
3. Cilantro
4. Spinach
5. Kale (full-size)

Our four secondary crops include:

1. Cucumbers
2. Carrots
3. Basil
4. Sugar snap peas

This list will surely change—possibly by the time you read this. Our mix of customers changes every season, and seed prices and other costs might alter what we decide to grow. It's important to stay nimble and not get too attached to any one product or customer. Of course, we haven't forgotten how to produce a wider range of crops. We are ready to grow them again if circumstances change. While we focus on a vital few, we still have the "trivial many"—or "useful many"—to fall back on.[1]

## THE VITAL FEW CUSTOMERS AT CLAY BOTTOM FARM

When we moved, we made the difficult decision to stop selling to some long-time customers. Before our downsizing, we sold to about a dozen restaurants, fifty CSA customers, a local grocery co-op, and hundreds of individual customers at farmers markets. With kids, we couldn't continue to do it all.

We decided to drop three of our seven restaurants because they were either too far away or because they ordered in low volumes. We told them that they could still pick up from our farm, but that we wouldn't be able to deliver to them anymore. We decided to pare back our sales at the farmers market, selling only in the summer months and not year-round. We paused our CSA program.

## How to Set Prices with Inflation

The best approach to pricing during a period of inflation is to stay current—research the market often and make frequent adjustments. Don't get left behind.

We adjusted prices several times during 2022 to stay current. The key was open, frank, and frequent communication with buyers. Every month or so, we asked chefs how our prices compared to other food distributors. When we delivered food to the co-op, we checked in with the buyer to ask what larger wholesalers were charging for similar products, and we tried to match those prices. In our experience, most buyers are happy to share prices in an effort to support local growers.

We still go back and forth about where and to whom we should sell our food. We're not sure we made the right call in every case. However, having fewer customers means less time driving, less time collecting orders, and less time processing payments. Also, we have deeper relationships with each customer and provide better service than we could before. In part because of the deeper relationships, our sales per customer have crept up, which is part of the reason we earn as much as before.

## Essentializing through the Pandemic

On March 22, 2020, I was in the greenhouse transplanting tomatoes. It was a sunny day, and I felt lucky to work in a warm space, able to work with my

I was in the middle of planting tomatoes on the day restaurants were closed due to the pandemic. We lost most of our accounts overnight. We quickly planted new crops and found new customers.

Clay Bottom Farm assistant farm manager Nicole Craig with CSA items ready for delivery. During the pandemic, we designed a no-waste CSA where food was dropped into coolers that customers set on their porches, with no packaging involved.

hands, while other people toiled away in their offices. By noon, I'd planted about 300 tomatoes.

Then I heard the greenhouse door open. Rachel ran in and told me, "You've got to stop planting tomatoes." Indiana's governor had just issued his weekly radio address, and he called on all restaurants to close due to the pandemic.

We went into the house to brainstorm what to do. We'd already invested time and money into the tomato plants, which we were growing almost exclusively for restaurants. Restaurant sales at the time made up more than half of our business.

What now? Our vital core, so to speak, had just dissolved.

The chefs, of course, had no idea when they might reopen. We decided not to count on them, at least for the spring and summer.

Nicole, Clay Bottom Farm's assistant farm manager, suggested talking with folks in her neighborhood to see if they might be interested in receiving the food we'd already planted for restaurants. Almost all of them said yes without hesitation. So, we quickly organized a direct-delivery, no-contact version of a CSA. To supplement the items we'd planted for restaurants, we seeded turnips, radishes, edamame soybeans, and green onions, sometimes tucking these crops underneath tomatoes we'd already planted.

We asked customers to put coolers on their porches, and each week Nicole drove our food around town, dropping carrots, peas, and tomatoes directly into the coolers, with no plastic involved. (We asked customers to supply their own containers for baby greens.)

Nicole also set up a bulk sales email list of customers close to our farm who might buy tomatoes and cucumbers and other items in large quantities for preserving. This list of customers became essential in August and September, when we had nowhere else to go with a lot of peak-season crops.

We've since refocused on restaurant sales again. But we learned a lesson: Farm with a vital core, but be ready to quickly change course. Thankfully, we could lean on local customers to help us through economic turbulence. Another way to put it: Focus on the 20 percent, but don't forget about the 80 percent. It might come in useful someday.

# A Cozybear in Athens, Georgia

"It's cool to say no."

Deijhon Yearby, at 27 years old, runs a successful nursery and market garden in Athens, Georgia, where he was born and raised. He has an essentialist's mindset, a knack for separating vital work from trivial work—for working on the right things at the right time to propel his business to the next level.

Yearby's path to farming started with football. He was an offensive lineman for the Clarke Central High School Gladiators, a top-notch team. In 2009, they were voted number two in the state in an AP poll; in 2011, when Yearby first played for the JV team, they set a record for number of games won.

Yearby was one of the better players. He moved up to varsity in his freshman year. He trained hard. By his junior year, he weighed 285 pounds. That year, he received five college scholarship offers. His future seemed set.

Then his life went sideways. "I was blocking four people. I had my arms stretched out to contain them," he told me. The play was botched, and the four people all fell backward, on top of Yearby's knee, tearing his ACL. Doctors told him his football days were done.

Needing to pivot his career choice, he remembered his tenth-grade horticulture class. "I enjoyed growing plants inside of a greenhouse," he told me. One of his coaches happened to be the ag teacher. He told Yearby,

Deijhon Yearby. Photo courtesy of Kristin Karch.

"Hey, remember when you sat in my class and you enjoyed talking about plants and vegetables? You can make that into a career."

Yearby started to work at Young Urban Farmers, a program of the Athens Land Trust, during high school. He prepared growing beds, seeded, watered, pulled weeds, and harvested vegetables. He took a job as assistant farm manager at the Williams Farm, an incubator farm, where he led workshops for farmers. He started his own podcast in which he gave out farming advice, and he became a leader that other young farmers look up to.

After high school, Yearby matriculated at Athens Technical College, earning a degree in agriculture. In 2019, he started his own farm.

## A FOCUSED FARMER

Pareto emphasized the unequal distribution of effort and reward. Some efforts always pay off better than others. Yearby has an instinct for pursuing efforts that pay off.

For example, after graduating from Athens Tech, friends suggested he work for a nursery to gain experience and save money. "But if I did that, I would spend 10, 15 years working for somebody else and then I would never do it," he told me. "Why not jump in while I'm young? I've seen people who, when they're in their forties, sneeze and their back goes out. I don't want to be in that situation, trying to start a business when I'm too old."

Instead, he started a market garden to build skills and save money for a nursery greenhouse. In his first season, he tried to do too much—he grew too many crops and tried to sell in too many places—but he quickly adjusted. He more carefully chose, the next year, what crops to grow and which markets to sell at, rather than jump at every opportunity.

To do this, he carefully tracked his customers. He used his farmers market booth to start an email list of his most dedicated customers. That list became a lifeline during the pandemic. When other farms shut down—because markets closed—Yearby sold more than ever through his own online ordering system.

Now, Yearby has the nursery business that he always wanted. He spends his days potting up ajuga, wormwood, oregano, lavender. He sticks to a tight budget, and business is growing steadily. Again, he shrewdly chooses crops and customers. He uses the same email list and online ordering system. "Everything I learned business-wise with the market garden, I transitioned over to the nursery," he said.

*A lot of farmers don't have business savvy. They say, "Oh, if I show up and put it on the table, people are going to buy it. We*

*put in hard work, and we deserve this price for it." I get that.*
*But you have to start with what the customer wants and stay*
*focused on that.*

Yearby told me he runs his nursery by spending a lot of time deciding what *not* to do, that his success is mostly the result of being shrewd and selective, of saying no to bad opportunities.

*It's cool to say no. It's hard but cool. I know it's hard because,*
*in the beginning, I would say yes to everything. Now I choose*
*wisely: No to that one. Yes to this one. In the beginning, I was*
*doing everything. There's a new market Tuesday at eight*
*o'clock? I was there. Now, if it doesn't make business sense, I*
*don't do it. I can make that choice now.*

# CHAPTER 4
# Simplify Fieldwork

*Any intelligent fool can make things bigger and
more complex. It takes a touch of genius—and a lot
of courage—to move in the opposite direction.*

—E. F. SCHUMACHER

Our next task was to apply Pareto's observation to our fieldwork. Which tasks were truly essential? Which ones should we weed out, so to speak? If we wanted to get small, to work fewer hours, then hard decisions would need to be made. We needed to simplify.

As we designed fieldwork systems on our new farm, we simplified in two primary ways. First, we reduced our field tools from dozens to seven. Second, we implemented the no-till, deep-mulch system, reducing field preparation time by more than half while improving crop quality. Chapter 8 shows our tools and how we use them, and chapter 6 covers the deep-mulch system in detail. In this chapter, I describe the principles behind our new growing systems. There is no uniform path to simplicity, but here are concepts that work well for us.

1. Use 5S to organize small spaces.
2. Standardize work.
3. Farm with as few tools as possible.
4. Farm like a tree: Stop working so hard.
5. Find waste and replace it.
6. Prioritize the simple-and-productive quadrant.

# 5S to Organize Small Spaces

Lean practitioners use a system called 5S, developed by managers at Toyota, to keep workspaces organized and ready for production. Each S represents a practice that improves flow—sort, set in order, shine, standardize, and sustain. I wrote about 5S at length in *The Lean Farm*; here I'll present additional 5S tools we've found to be useful in the unique context of a tiny farm.

On tiny farms, even a bit of untidiness can grind work to a halt. On larger farms, there is an allowance for letting things go—a few tools or row covers or tarps can be left out of place and work can probably still go on. On a tiny farm, however, you need to get rid of as much as you can, keep spaces tidy, and clean often.

## 1. SORT: GET RID OF ONE THING EVERY DAY.

Sorting, the first 5S step, is a discernment process where touch is the key metric. Which tools get routinely touched and put to work? Which ones collect dust? If tools aren't put to frequent use, adding value to your products, it's time to say goodbye.

Sorting can be ruthless. Unused tools and supplies should not go into a corner or under a table. (This is called squirreling, not sorting.) They should be physically removed from your production area. The only tools and supplies left behind should be absolutely essential.

In a tiny-farm context, a powerful sorting practice is to get rid of one item every day. Our property is not always clean; in fact, it is often cluttered. It seems like items we don't really need magically appear. But we have found that sorting a little bit every day helps us take back control. This takes dedication because, at least for us, it is easier to add than to subtract.

One benefit of daily sorting is that you won't need a large storage room. If you sort once a year, then you need a place to put things until sorting time. As stated, we built just enough space on our farm. By design, there is little room for overflow.

A second benefit: Everyday sorting sharpens our eyes. If we only sorted once per year, we would lose focus. We would start to tolerate an infiltration of useless items. Everyday sorting has us on the lookout all the time for what doesn't belong.

In practice, we put a recycling bin, trash bins, and a sorting table next to our processing room door. We place on the table items that have crept into our

## If You Keep Extras, Set Limits

In reality, there are good reasons to sometimes hang onto things—even if they're not truly essential—as long as they are dealt with properly. I enjoy visiting farms with spare tractors up on jacks and with scrap metal piles for projects and tinkering. My neighbor Jon, who passed away during the writing of this book, was one of these farmers. I looked up to Jon because he could fix anything and because he was generous with his spare parts. Nothing gave him more joy than seeing something he'd saved for years put to use on our farm. To build our greenhouses, we used salvaged electrical service panels, chunks of old cement sidewalks, and copper wiring, all courtesy of Jon.

It is important, though, to establish clear boundaries between your production space and storage spaces. Piles of scrap metal might be fine behind the barn, but they don't belong in your greenhouse or processing room. We have an old barn a few hundred yards from our growing area where we store spare trellising gear, old air conditioners, and scrap lumber. My kids enjoy rummaging through the barn and pulling out stuff to build forts, bicycle ramps, and other creations. But in our small production space, we keep only the essentials.

We built a lean-to, attached to a greenhouse, for row covers and shade cloth, where these supplies are easy to access. The rule is to store tools and supplies as close as possible to their points of use and in plain view.

production area that don't belong there, like toys, soccer balls, fishing poles, and leftover lumber. Every week or so, we haul these things out.

## 2. SET IN ORDER: MAKE EVERY TOOL VISIBLE.

Setting in order means that every tool should have a home. Again, the practice is ruthless. A tool should either be in your hands or in its home. No wandering permitted.

The key point when setting tools in order is to store them as close as possible to their points of use. A central tool storage area is rarely a good idea. It's better to spread tools out where you can quickly grab them.

On a tiny farm, I add the rule: *Make every tool visible.* On our 1/3-acre farm, we can see all of our field tools—they are not stowed away in sheds. We use a beam under our spray station to hang shovels and hoes, next to a path leading to the greenhouses and field plots. We use magnet strips to hang knives and pruners on our greenhouses, next to the entrances. This system puts every tool within sight. In the wintertime, we move all of our tools inside of our greenhouses to protect them from the elements and to keep them nearby for winter work. We also keep supplies fully visible, storing clamshell containers, row covers, stakes, and other necessities in plain view.

## 3. SHINE: NO DIRTY CORNERS.

The third step is to deep clean your spaces and make sure they are well lit. Of course, this is easier to do after you've removed nonessential items.

Cleanliness is most important in washing and food preparation areas. High-lumen LED light fixtures brighten our work and make cleaning easier to do in these areas. We also covered the walls in these spaces with sheet metal so that we could powerwash the walls.

We even clean in the greenhouses. We remove weeds, of course, but we also sweep and hose the sidewalks inside and out, and once a year we

After tidying, the next step is to deep clean. A clean farm promotes quality work.

powerwash the polycarbonate end walls. This keeps the space tidy and more hygienic for plants.

The new rule I add here: *No dirty corners.* It can be easy to neglect corners. Tools, unused supplies, and stuff without a home seem to be magnetically drawn to them. When cleaning in our processing room and in our greenhouses, we sometimes start with the corners, to make sure they get done, and then work inward. I've found that washed corners are an indicator of overall cleanliness.

## 4. STANDARDIZE: CLEAN AS YOU GO.

Standardizing under 5S means to set up routines so that sorting, setting in order, and shining happen routinely, not sporadically.

We keep tools handy for sorting, setting in order, and cleaning to encourage daily use. For example, under our spray station porch, we hung a hose reel so that we could quickly wash down tools and totes and neatly store the hose. We hung an air compressor reel with constant pressurized air within easy reach on the same porch. We use the hose to blow dirt off tools and ourselves (and of course to fill up tires). On a post under the porch, we hung a battery-powered leaf blower that we use to clean the porch floor.

In the processing room, we hung squeegees and brooms where we can quickly grab them.

For tiny farms, I add the rule: *Shine as you go.* My inspiration for this phrase was Sally Schmitt, the founding chef at the French Laundry restaurant, who wrote in her memoir/cookbook, *Six California Kitchens*, that her motto was to "work clean, work neat, and work carefully." Schmitt said that the most efficient way to cook is to clean as you go—to wash knives as soon as you are done cutting with them, to rinse out pots and pans as soon as you are done with them, and so on.

Cleaning as you go saves motion. Consider a mixing bowl. In most kitchens, once the frosting has been whipped, the mixing bowl probably is set on the counter to be cleaned up later. In Schmitt's kitchen, each cooking session started with filling a large bowl with hot soapy water. When a mixing bowl was emptied, it went straight to the wash water. She saved at least two steps with each dish because she didn't have to set the dish down and pick it up again to clean it.

Also, orderliness promotes efficiency. When dishes pile up, you have to work around them. If your counter is small, your growing pile of dishes will eventually cover another item that you desperately need to find. There is a psychological component as well. In cluttered workspaces, it can be difficult to focus. Tidy spaces promote attention to the work at hand.

Farms are no different from kitchens in this sense. In the early years of our farm, we thought we were too busy to shine as we went. Instead, we often left tools where we last used them and row covers at the edge of our fields. As a result, by the end of the season, we too frequently spent as much time looking for tools as we did using them. When we started to clean every day, our work went more quickly and was more enjoyable to do.

Here are specific clean-as-you-go tips, adapted from Schmitt:

- Start the day clean. Don't have things piled up to be washed from the day or night before. This gives the day a feeling of a fresh start.
- Keep hoses, brushes, and other items needed for cleaning handy.
- Wash totes, crates, and garden tools right after using them.
- Don't leave a knife on the counter or otherwise out of place. Use it, wash it, and return it to its place. Your knives will stay sharper if you do this, and it is safer.
- Leave harvest containers out to dry; air drying is more sanitary than towel drying.
- Frequently wipe down counters, scales, and other food-contact surfaces. This saves a lot of hard cleaning later.

## 5. SUSTAIN: LINK NEW ROUTINES TO EXISTING ROUTINES.

The final 5S step is to integrate organizing into your farm's culture. *Kata*, which translates to "form," is a Japanese word used by lean practitioners to describe this concept. In martial arts, kata refers to movements that are practiced over and over again until they are reflexes. Sorting, setting in order, and cleaning can become reflexes, too.

The best way to make progress in a new routine is to link it to an existing routine. For example, I recently decided to take up meditating for a few minutes each day. At first, I had trouble finding the right time to do it. I thought it would be easy to slip it into my day, but it wasn't, and many days, I just didn't do it.

Then I started to "stack" my meditation onto my existing exercise routine. Most afternoons, I do a short workout. By adding meditation to the end of each workout, I no longer have to think about when to do it; it happens as part of a routine. This is a technique I learned from management teacher S. J. Scott in his book *Habit Stacking*.[1]

Likewise with 5S tasks. Each of these tasks can be linked to other farmwork. For example, my weekly kata, so to speak, is to clear the sorting table, while others on the crew begin washing greens. This links sorting to the task of washing greens. There are certainly days that get missed, but the table stays much cleaner than before we instituted the practice.

Another example is sanitizing the processing room. For a time, it was unclear who this job belonged to. Sometimes we just skipped cleaning. We decided to link the task to greens washing. Each time greens are washed, the same person sanitizes the processing room. Cleaning isn't a stand-alone assignment—it's linked to another one.

# Standardize Work

This practice of routinizing tasks can serve other areas of the farm, too. Standardized work is a practice at the heart of Toyota's lean system, and it really just means that work should be broken up into manageable small chunks done the same way every time, according to a standard. In other words: to tame chaos, reduce variability.

This ideal can be hard to achieve on diversified vegetable farms, which seem to lend themselves to chaos. Changing weather, infuriating insects, and unforeseen plant diseases disrupt the most carefully made plans. Still, it is possible to standardize a lot of farmwork.

Standard work means jobs are performed to the same standard every time, as much as practical. Photo courtesy of Adam Derstine.

On the surface, standardizing work might seem to sap creativity. Who wants to do the same thing the same way every time? To a degree, this is a danger. Especially in large factories, work can be monotonous and mindless, too standardized. But when it is implemented well, standardized work paradoxically increases creativity and innovation.

Let me explain. First, standard work increases workers' confidence in what they are doing. We see this on our farm. When workers know how to complete their tasks, they are happier, more engaged, and more motivated to perform quality work than when a task is confusing. There is positive psychology at work here. When we know what to expect, we relax and focus better.

Second, standardizing *routine* work—getting it done more efficiently—frees up time for *creative* work. My kids might put it this way: If we get our chores done faster, we can play ball sooner.

We have seven core standardized tasks on our micro farm.

1. Preparing growing beds
2. Seeding paperpot trays
3. Direct-seeding greens

4. Harvesting tomatoes
5. Washing greens
6. Packaging tomatoes
7. Delivering produce

These are tasks that are done in essentially the same way every time. As a farm team, we have developed a mastery of these tasks. We get them done quickly, thus freeing up time to problem-solve more complicated aspects of the farm, like installing an innovative ridge vent in a greenhouse or redesigning a germination chamber.

With standard work procedures, training is also easier and faster. The steps for each of our seven core tasks are clear, so I can teach workers quickly, typically in a few minutes. In fact, these jobs are so streamlined that workers often cross-train. This relieves me of the burden of training every employee on every job. For example, in 2022, at least four new workers or volunteers were trained on how to wash and package greens, but I did none of the training myself.

## No Yellow Papers: Keep Changing the Standard

Standard work does not mean static work. On the contrary, standards should always be changing. In the early days at Toyota, standards were written down on white sheets of paper that were hung up in the workplace. It took about one month for the papers to turn a bit yellow, due to light exposure. Taiichi Ohno, a manager at Toyota, used to decry these yellow papers, as they indicated mental laziness, that standards were not being updated fast enough.

This work of improving standards is a job for everyone. John Shook, who worked for Toyota in Japan for 11 years, says that "the Toyota Way is a socio-technical system on steroids," meaning that it "brings together people (social)

and process (technical) to bear on purpose."[2] People are encouraged to actively engage with their work, not just complete their jobs. Or as Ohno put it:

*The Toyota style is not to create results by working hard. It is a system that says there is no limit to people's creativity. People don't go to Toyota to "work"; they go there to "think."*[3]

Working together to improve standards is the ideal way to engage workers. Whenever we train new workers, we tell them that part of their work is to find a better way to get the job done. This puts all of us on the same page, as it were, trying to improve every process every time.

Another benefit for managers: less variety of output. With standard work, I can be confident that results will match what I want, no matter who does the job, because the process does not change even if the people change. When harvesting tomatoes, for instance, all workers use the same set of pruners and the same harvesting bucket; they all follow the same grading standards; and they put the same label on tomato boxes in the same place. This predictability is a comfort to employee and employer alike.

## Farm with as Few Tools as Possible

Humans have been growing food for more than 10,000 years with scant tools: a grubbing hoe to work the earth, a knife for harvesting, and perhaps a basket to hold the harvest. Only since the industrial revolution has farming turned into a tool frenzy. We are not afraid to bring a new tool to the farm, but new tools need to meet a high standard of productivity.

Every season our goal is to farm with fewer tools than in the previous season. Let me list the advantages we have seen from using so few tools:

1. *Less training.* I can train new workers on all our tools in just a few hours. Many of our workers are college students who stay with us just a few months. It is handy, from a management perspective, to know that everyone on the team—even short-termers—can use any tool on the farm.
2. *Tools not lost as often.* We store our seven field tools in one place, lined up so we can easily see if one is missing. While a tool still does occasionally grow legs and wander, there is less chance of that happening if tools are small in number to begin with.
3. *Fewer choices to make.* Barack Obama once told reporter Michael Lewis, "You'll see I wear only gray or blue suits. I'm trying to pare down decisions. . . . I don't want to make decisions about what I'm eating or wearing. Because I have too many other decisions to make."[4] The same principle applies on a farm—fewer tools mean fewer decisions and more brain space for what matters.
4. *Lower cost.* Simply put, fewer tools mean less money spent on tools. In our case, we arrived at fewer tools by a process of subtraction. We'd overbought and then pared down. If you are starting out, you can avoid this problem by choosing more wisely from the beginning.
5. *Appreciation for each tool.* Looking at our tools is sometimes like looking at a photo album. I get nostalgic remembering projects the

tools helped us accomplish, and I think of interns and workers who used the tools in the past. Also, with fewer tools, we can afford to invest in better ones.

6. *Less storage space required.* Our vital seven tools require about 10 square feet of storage space. By contrast, I've been to many farms that seem to be designed around storage spaces for an overabundance of tools.

7. *More satisfaction.* We enjoy our work more when our "stuff" isn't center stage. With fewer tools, we are more focused on our work and less distracted by managing our things.

We accomplish most work with just seven field tools. Photo courtesy of Caleb Mast Photography.

Like all principles, this one can also be taken too far. Sometimes, adding another tool—the right tool—can make all the difference. A recent example for us is our battery-powered leaf blower. For years we broomed our porches and sidewalks and swept the cobwebs from corners by hand, resistant to adding another tool. Last year we invested in a leaf blower, which completes the job in a fraction of the time and does it more thoroughly. The best approach with tools, though, is to be spare, adding new ones rarely and only after carefully thinking through the pros and cons.

## Farm Like a Tree: Stop Working So Hard

A single tree, though it never moves, can be astoundingly productive. Mature pecan and chestnut trees produce about 100 pounds of nuts in a single season; a healthy apple tree will grow 15 bushels of apples in a year; a sugar maple tree, in a good season, will deliver 20 gallons of sap. How do they do it? In the past few years, we have planted hundreds of trees on our farm, reforesting areas that we are not using to grow crops. I've observed and thought a lot about trees recently. Here are four observations that I think help explain their productivity.

## Use Technology Designed for Small

In his travels to Burma, E. F. Schumacher saw Western mass-production tools imposed on villages where the tools didn't fit. He was initially sent to Burma to encourage the use of these new technologies, but he quickly realized that the Burmese "need no advice: as long as they don't fall for this or that piece of nonsense from the West."[5]

When he returned to England, Schumacher championed "tools with a human face," or appropriate technology: tools designed to fit the people who use them, with input from the user. Tools, he said, should be a "democratic or people's technology ... to which everybody can gain admittance and which is not reserved to those already rich and powerful."[6]

Sadly, in the world of small farming, many tools do not seem to have people in mind. Too many designers never consult farmers. For example, tractors and implements purportedly designed for compact farms are often unwieldy to navigate and too heavy, causing compaction. Farmers are forced to fit their farms around their tools when it should be the other way around.

When gathering tools for a new farm, I encourage you to seek out those designed with the micro farmer specifically in mind. I list many such tools in chapter 8. I've found that some of the best tools come from Korea and Japan, where agricultural engineers focus on designing

quality tools for tiny-scale farms. (Farms in these countries are almost all small-scale by US standards.) In the United States, the best-trained agricultural engineers are employed by John Deere and Caterpillar and serve the needs of megasize industrial farms.

Tools with a human face are purposely designed for operation on a small scale. The Japanese paperpot transplanter, pictured here, is an example of appropriate technology. Photo courtesy of Caleb Mast Photography.

## 1. TREES WORK AHEAD.

Trees can be so productive because they spend years preparing. The fruit and nut trees that we planted, for example, use their first few seasons to create a strong scaffold of branches before producing heavy loads of fruit. They prepare ahead for high productivity.

Similarly, we've learned that it is better to set up carefully before jumping into production. Before our season begins, we take stock, making sure we have ordered enough (but not too many) row covers, clamshell containers, tomato boxes, and other supplies that we will need in the months ahead. As I write this, in the midwinter, we are preparing to transplant tomatoes in a greenhouse in March. We are disinfecting trays, landscape fabric, and the germination chamber and taking inventory of labels, stakes, and other supplies necessary for the job.

We work ahead in the field, too. If we want to use a bed in the spring, we'll prepare it in the fall, adding compost months ahead of when the bed will be planted. (The compost itself was made at least a year ahead.) We have more time in the fall, so we are less rushed and more likely to do a better job. It is best not to farm in the moment, when you are prone to always playing catch-up. Think ahead and prepare.

## 2. TREES LEAVE PLANT MATTER IN PLACE: THEY COMPOST IN-SITU.

Trees understand the value of a dead leaf. A full-grown maple tree has about 500 square feet of leaves weighing about 500 pounds. The total chloroplast surface area of a single tree equals about 140 square miles—an impressive solar array. By fall, these leaves are full of carbon, nitrogen, phosphorus, potassium, calcium, and other nutrients.[7] It is a great tragedy that so many people feel compelled to throw leaves in the trash—or, worse yet, to burn them.

The trees we planted don't move their leaves. They just let them drop and decompose in-situ, in their original place, to feed themselves and organisms in the soil life. Many gardeners move dead plants out of the garden at the end of the season to clean up the garden and supposedly prevent

Our rule is to leave roots in the ground to feed the soil.

disease. For many years, because of what we were taught, we cleaned up in the fall like this, pulling plants, roots and all, and moving them to a compost heap.

Now we take a different approach. As much as practical, we let old plants decompose in place, where minerals from their leaves, stalks, and roots feed the soil, like a tree. The chapters ahead show high-productivity techniques for farming while leaving plants in place.

### 3. TREES PRACTICE QUIESCENCE.

Conventional wisdom says that trees in winter are dormant to conserve energy. Aboveground, this is mostly true, but new research shows that belowground is a different story.

Michael Snyder, writing for *Northern Woodlands* magazine, explains, "Tree roots seem to maintain a readiness to grow independent of the aboveground parts of the tree. That is, roots remain mostly inactive but can and do function and grow during winter months whenever soil temperatures are favorable, even if the air aboveground is brutally cold."

This is called *winter quiescence,* where roots are resting but ready. It is a survival tactic important for the health of the tree. "It is this trait that allows evergreens to absorb soil water and avoid winter desiccation in their needles," writes Snyder. Quiescence "allows all species, including deciduous hardwoods, the opportunity to expand their root systems in search of water and nutrients in advance of spring bud break."[8]

On a farm, winter is a great time to both rest and expand. Days are shorter, and there is less work to do. It's important to sleep more and work less. But the cold season is also the best time to explore new growing techniques, redesign tools, and visit other farms to learn new skills in preparation for the spring. Trees are a case study proving it is possible to rest and prepare at the same time.

## Find Waste and Replace It

*"Take care of the waste on the farm and turn it into useful channels" should be the slogan of every farmer.*

—George Washington Carver

The lean system can be boiled down to a simple dualism: Every activity on your farm either adds value to your products or it contributes a form of waste. There is no middle ground. The work of a lean practitioner is to scour an operation for the waste and then get rid of it. Or better yet, replace it with an activity that actually adds value.

Here is a field-management-level way to apply this "replace waste" mindset. Imagine you grew ten beds of vegetables and that eight beds succeeded—the crops grew to maturity and sold at a profit—while two failed—the crops died or went unsold. Let's say that each bed absorbed $1,000 in costs and that crops in each bed grossed $2,000 in revenue (again, I am exaggerating to make a point). That year, then, your costs were $10,000 and your gross income was $16,000.

Let's say your goal next season is to reach ten successful beds. There are two options. First is the growth model: You could work up more ground and plant more, assuming some beds will fail. In this case, let's say you decide to plant twelve instead of ten beds. Let's say that again two beds fail.

This growth model seems logical: If two beds fail, you still reach your goal of ten successful beds. You've soaked up waste like a sponge by doing more. You've used overproduction as a safety net. You've scaled up to absorb losses. This is classic "get big" thinking and is a traditional way to run a farm business.

But there's a false comfort here. First, because you produced more, your total costs went up. You had to invest more capital in your business. You probably had to purchase or rent more land. You needed to buy more seeds,

| | |
|---|---|
| 1 | Success |
| 2 | Success |
| 3 | Success |
| 5 | Success |
| 4 | Success |
| 6 | Success |
| 7 | Success |
| 8 | Fail |
| 9 | Fail |
| 10 | Success |

| | |
|---|---|
| 1 | Success |
| 2 | Success |
| 3 | Success |
| 5 | Success |
| 4 | Success |
| 6 | Success |
| 7 | Success |
| 8 | Fail |
| 9 | Success |
| 10 | Success |
| 11 | Success |
| 12 | Fail |

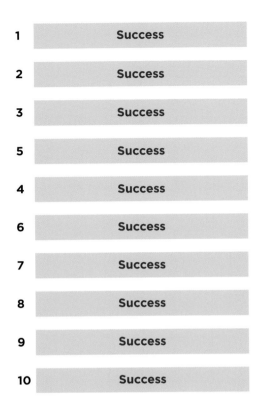

| 1 | Success |
| 2 | Success |
| 3 | Success |
| 5 | Success |
| 4 | Success |
| 6 | Success |
| 7 | Success |
| 8 | Success |
| 9 | Success |
| 10 | Success |

fertilizer, and compost. You had to work more hours in order to create the new growing beds and to seed, tend, and harvest more crops.

Consider another model.

Instead of adding on to your farm, replace the two beds of unsold/failed crops with two crops more likely to succeed, and tend them better. At the end of the season, as with the previous option, you will still produce ten successful beds—but at a lower cost. You buy fewer seeds, fertilizer, and compost. You don't need to till more ground. You reduce your work hours by almost 20 percent and earn a higher margin!

To be sure, the above models are simplistic. But the models demonstrate a powerful fact: By taking care of waste and turning it into useful channels, you achieve more production with less effort and cost. You do less but better. This way of running a business is countercultural. Every incentive exists to grow and do more and get bigger every year. But if you don't have access to endless capital and a gigantic tract of land, then the second model provides a brilliant way to compete.

Here is a real-life example. On our new smaller farm, we have approximately fifty growing beds. Most beds are used to grow crops twice each year, for a total average of around a hundred crop cycles per season. In our first season at the new farm, approximately twenty beds failed, mostly because we had new land and microclimates to get used to. Also, we were busy, building infrastructure at the same time, and things got overlooked. Specific beds that failed were as follows:

- Green beans: We planted two beds more than we could sell.
- Tomatoes: A row fell over because we used wooden stakes that were too old.
- Peppers: A wet spring caused plants to rot.
- Eggplants: We planted one bed more than we could sell.
- Lettuce: Four beds were eaten by deer.
- Zucchini: We planted too many and did not harvest them in time.

## George Washington Carver and the Tuskegee Model of Replacing Waste

In popular media, George Washington Carver is known as the "Peanut Man" because he invented more than 300 uses for peanuts and pioneered innovative ways to grow them. He traveled the country preaching the peanut gospel, as it were. His efforts worked. He helped to revitalize the rural South in the early 1900s, because growing peanuts became a way for farmers to replace cotton.

But "Peanut Man" does not quite capture all of Carver. Carver's project, at its heart, was to create opportunities in agriculture for poor farmers who had limited tools and capital. In his 1905 pamphlet, *How to Build Up Worn Out Soils*, he writes, "The Tuskegee station has ... [kept] in mind the poor tenant farmer with one-horse equipment; so therefore, every operation performed has been within his reach."[9] Every experiment he performed had tiny-scale farmers in mind.

At the heart of Carver's teaching was the replace-waste concept. He wrote that farmers should "take care of the waste on the farm and turn it into useful channels." His pamphlets implored growers to turn crop residues into compost. He even advocated for composting human manure and using it to fertilize crops.[10] He taught methods for canning and preserving unsold crops. He modeled the waste-free thinking in his lifestyle: He owned one wool suit throughout most of his life, gave money away generously, and kept

George Washington Carver at the Tuskegee Normal and Industrial Institute, Experiment Station, examining crops, circa 1906. Carver and his colleagues focused their work on methods geared toward tiny-scale farmers. Photo from the Tuskegee University Archives, Tuskegee University.

few possessions so that he could focus his energy on his work. To build his research laboratory, he wrote that he "went to the trash pile at Tuskegee Institute" and pulled out "bottles, old fruit jars, and any other thing I found I could use."[11] Carver's talent wasn't just to invent, it was to see the value and opportunity in the things other people considered to be waste.

- Carrots: One bed was mostly eaten by rabbits.
- Edamame: One bed was eaten by deer.

As stated earlier, I was nervous that our decision to farm on a smaller footprint wouldn't pay off, that we wouldn't produce enough to support our family. This first season seemed to confirm my fears. At the end of the season, Rachel and I discussed the option of plowing up more land and adding more growing beds.

Instead, we decided to step back. We analyzed our failures until we found solutions to each problem. We built better fences to keep out pests. We found better ways to stake tomatoes—with metal T-posts in between wooden posts. If no economically realistic solutions existed, we replaced the crop. Our farm is far from perfect, but as of this writing, now more than 95 percent of our beds produce crops that sell.

# Prioritize the Simple-and-Productive Quadrant

We might summarize this chapter by saying that the way to simplify a farm is to create tidy spaces with 5S, standardize work, farm with treelike efficiency, choose a few good (that is, appropriate and human-scale) tools, and ruthlessly replace waste with value.

One thread ties these practices together: All require making an overwhelming number of decisions. There are so many tools to choose from, so many ways to grow crops, so many types of waste on a farm, that it can be confusing to decide how to move forward.

All solutions to a given problem can be broken down into these four types:

1. Complex and productive
2. Simple and productive
3. Complex and inefficient
4. Simple and inefficient

Ideally, you find approaches that are both simpler and more productive (#2). These methods are low cost to do but get a lot done. For example, we opted for the deep-mulch method at our new farm, instead of cover cropping and tilling, to build soils, fertilize, and smother weeds. This approach accomplishes all of these tasks in one go. There is no specialized technology or technique involved. The only tools required are a shovel

The best way to propel your farm is to persistently simplify everything you do.
Photo courtesy of Caleb Mast Photography.

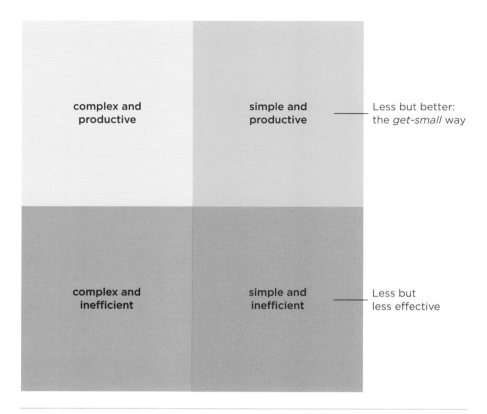

|                          |                       |                                       |
| ------------------------ | --------------------- | ------------------------------------- |
| complex and productive   | simple and productive | Less but better: the *get-small* way  |
| complex and inefficient  | simple and inefficient | Less but less effective              |

Solutions to production problems fall into four categories. For a minimalist micro farm, steer toward solutions in the upper-right quadrant: simple and productive.

and a rake. It can be done quickly and efficiently. It is a win-win for simplicity and for productivity.

To be sure, there are exceptions. For example, when harvesting baby greens, the simplest solution would be to cut greens with a knife; instead, we use a mechanical greens harvester, a small battery-powered machine. But in most cases, the best solution to a problem is also the simplest.

Simple and productive solutions require creative thought to develop and sometimes a willingness to try things no one else is doing. As E. F. Schumacher writes, "Any third-rate engineer or researcher can increase complexity; but it takes a certain flair of real insight to make things simple again."[12]

The pursuit of simplicity is essential to getting small. At its core, a productive small farm is powered by straightforward approaches, not by complexity or technology. This is how we are able to earn a comfortable living on ⅓ acre of land.

# Tiny Giants in Kalamazoo

*The best way to manage tools on a tiny farm
is to get rid of as many of them as you can.*

Ben Brown and Julia Whitney Brown use their ⅓-acre backyard in Kalamazoo, Michigan, to grow vegetables for a living. Their operation, Tiny Giant Farm, is tidy, tiny, and productive. They fuel their success by persistently simplifying everything that they do.

## A CONSIDERED FARM LIFE

Ben Brown went to college to be a sociologist, but two years in, he realized he'd made a mistake. He couldn't bear to sit down all day. He needed a more active career.

Ben Brown and Julia Whitney Brown at their farmers market booth. Photo courtesy of Tiny Giant Farm.

So, he decided to become a rock star. He formed several bands and recorded an album for sale on iTunes and Spotify. But the financial pressures were too great, and endlessly traveling around the country was hard on his body.

He changed gears again. He enrolled in an ag program at Sterling College, a tiny school in northern Vermont, where he learned how to raise sheep, grow a market garden, and perform maintenance on a chain saw.

In Vermont, he also met Julia, who had just finished a ceramics program in Illinois. Julia told me that she realized in Vermont that "pottery could fit in really well with agriculture, with a shared direction—Ben a farmer and me a potter: We both work from the earth and literally bring our gifts to the table."

After a stint in Nashville, where Ben worked as farm manager at a nonprofit, he sent me an email asking if I needed a farm hand for a few days a week while he and Julia started Tiny Giant Farm, an hour north of us. With his credentials, I eagerly said yes. Ben worked with us during the first season on our new micro farm, then he went on to work full-time at Tiny Giant Farm.

## FARMING SIMPLIFIED

Ben makes day-to-day decisions in the gardens and performs fieldwork with a part-time staff person. Julia helps sell at a weekly farmers market—just 3 miles from the farm—and makes management decisions with Ben. Julia is also now a graduate student in a ceramics program. They have a daughter, Naomi, who is two years old as I write this.

Ben and Julia are always looking for less-but-better solutions. One example: tool selection. In our interview, Ben quickly rattled off field tools that he regularly uses:

1. stirrup hoes (two of them)
2. bed rake
3. tine rake
4. hoss wheel hoe with sweeps
5. paperpot transplanter
6. BCS tiller (they only use the highest till setting)
7. wheelbarrow
8. shovel
9. digging fork
10. mist sprayer
11. greens harvester

Overhead view of Tiny Giant Farm. Photo courtesy of Tiny Giant Farm.

"That's really it," Ben said. He has tried hundreds of tools in his career, but he is more animated about discarding tools—finding a just-right number of tools—than about acquiring new ones. "The best way to manage tools on a tiny farm is to get rid of as many of them as you can," he told me.

Another way Ben simplifies is by working ahead. For example, he weeds pathways on a regular schedule with a wheel hoe equipped with fixed-blade sweeps, cultivating even before he sees weeds. "It's essential, on this tiny scale, to keep things clean." It is much faster to prevent weeds by weeding ahead than to dispose of them after they appear.

Ben and Julia have standardized almost every element of the farm. They divided growing spaces into six plots, each 32′ × 50′. Each plot has eight growing beds that are 30″ wide. "This way we can use the same irrigation system, the same row covers, the same silage tarps on everything," Ben said. "That's been essential to our efficiency." They've also simplified plant spacings. "We grow almost everything in the 6″ paperpot chains," he said. The only exceptions are beets, which grow in 4″ chains, and scallions, which grow in 2″ chains.

"We grow all of our tomatoes with the same 12″ spacing, with two lines of drip tape, all with the Qlipr system, all with two leaders, and all on the same trellising schedule," he said.

Washing and packing used to be a jumbled mess because their space—a one-stall garage—was so small. "We had to standardize," he said. Now they use one size of container, 18-gallon Rubbermaid totes, that they harvest and pack into. "Each tote holds 23 units, no matter what's in it—23 carrots or bags of greens or whatever. It's all the same."

As you might expect, Ben and Julia even simplified their farmers market booth. They sell almost everything in bags or boxes so there is no weighing, using uniform pricing: $4 per item, two items for $7, or three items for $10.

All of these efforts, of course, save costs and increase productivity. But Ben and Julia pursue simplicity primarily for another reason: to spend more time with Naomi. Julia told me:

> *Our goal for saving time is not to make as much money as possible—or to be perfectionistic in our processes. Rather it's about nurturing each other. . . . A plant doesn't grow alone—it requires healthy soil, sunlight, warmth, water, and a place to be rooted. This is the metaphor we use to structure our work and our life.*

Ben added:

> *Now, it's five o'clock and I'm done. It doesn't matter until tomorrow. Having Naomi has made me a better business owner and farmer, because having a kid establishes certain boundaries. You've got to be creative to fit everything in the boundary. I want to be able to hang out with the important people in my life without being completely exhausted.*

# CHAPTER 5

# Localize: The Practice of *Swadeshi*

*We must manufacture our own cloth and, at the present moment, only by hand-spinning and hand-weaving.*

—MAHATMA GANDHI

On July 31, 1921, Mahatma Gandhi and his supporters piled 150,000 English-made cloths, including expensive saris, into a circle 20 feet tall at the Elphinstone Mill compound in Mumbai, India. Before a crowd of more than 10,000 people, Gandhi declared, "We are purifying ourselves by discarding foreign cloth, which is the badge of our slavery."[1] Organizers lit the pile on fire, and Gandhi took a pledge to boycott foreign goods, launching the Swadeshi movement.

*Swadeshi* is a conjunction of two Sanskrit words for "self" and "country." Swadeshi came to stand for all things made in India by and for Indians, for self-sufficiency in the face of British rule.

But Swadeshi for Gandhi wasn't a jingoistic rallying cry. He defined Swadeshi as the "spirit in us which restricts us to the use and service of our immediate surroundings to the exclusion of the more remote."[2] He aspired for villages to support themselves by growing their own food, building houses out of local materials, and weaving their own clothes, as a means of mutual betterment and, ultimately, spiritual enlightenment. After the rally in Mumbai, Gandhi

Gandhi frequently spun yarn with his charkha while speaking in front of crowds. Photo courtesy of Wikimedia.

lived in a self-sustaining community, wore homespun clothes, ate simple meals, and maintained few possessions, to embody the Swadeshi lifestyle.

Gandhi was the first person to use the word *Swadeshi*. There is no English language equivalent. However, I think that the verb "to localize" points to the spirit of Swadeshi. Dictionary.com defines localizing as "to gather, collect, or concentrate in one locality."[3] Oxford Languages says that to localize is to "restrict something to a particular place ... adapt (something) to meet the requirements of a particular area."[4] Gandhi wanted an independent India rooted in local self-reliance, what he called "enlightened self-control and self-development"—a network of localized villages adapted to their place.[5]

Small businesses, like farms, should pursue Swadeshi, Gandhi said—through innovation. Local industries must continually improve their methods to produce goods efficiently and of persistently higher quality, to make local goods attractive compared to shipped-in counterparts. In his own words: "I should use only things that are produced by my immediate neighbours and serve those industries by making them efficient and complete where they are found lacking."[6] Gandhi himself invented a double-wheel charkha, or spinning wheel, that spun yarn faster and with more control with a smaller-size wheel. The charkha was a symbol of resistance to British rule, generally. But importantly to Gandhi, it was also a symbol of pride in village-based economics.

He envisioned households in a village freely exchanging goods and services as needs arose, not necessarily based on market forces. In this way, "every village of India will almost be a self-supporting and self-contained unit, exchanging only such necessary commodities with other villages where they are not locally producible."[7]

In chapter 1, I described how E. F. Schumacher in Burma became disenchanted with Western economic models of growth without boundaries. In Gandhi, Schumacher found an alternative—one that put people first. Swadeshi was so practical, said Schumacher, that even communities within rich societies should adopt its principles. In his introduction to *Small Is Beautiful*, Theodore Roszak writes,

> *Gandhi's economics, for all its lack of professional sophistication (or perhaps for that very reason) was nonetheless the product of a wise soul, one which shrewdly insisted on moderation, preservation, and gradualism. ... Gandhi's economics started (and finished) with people, with their need for strong morale and their desire to be self-determining.*[8]

Already, in our first full growing season at our new farm, we were beginning to reap the rewards of our move into town. Boundaries gave our work focus and energy. Better infrastructure and simpler growing systems made us more productive. It felt *good* to farm again.

The next step was to integrate into our local community, to try the Swadeshi model on our farm.

In our current economy, every incentive exists to look beyond the local community—for example, to pursue higher-paying and more reliable customers farther away. It takes intention and hard work to move in the opposite direction, as Schumacher might say. But the economic and social payoffs for a tiny farm are worth every effort. This chapter is about the strategies we use to sell all of our food within 1.5 miles from where it's grown.

## The Ancient Art of Weaving

To understand Swadeshi at a deeper level, think about a piece of thread. On its own, a piece of thread is relatively weak. It can be easily cut or maybe even pulled apart by hand. It is barely useful. One piece of thread can lift or pull only a tiny amount of weight. But when strands of thread are woven together, the result is a strong rope that can hoist thousands of pounds and attach giant objects together, or a cloth that can last generations. Gandhi understood the powerful symbolism of threads woven together, part of the reason he used local homespun cloth as a metaphor for the strength of local villages.

To localize is to weave. First, weaving out—making your food (and yourselves) available to the community that surrounds the farm. Second, weaving in—bringing the community onto the farm. Thus, woven out and in, bound together, farm and community become integrated. There are things schools, governments, and others can do to weave farms and local communities together. This chapter, however, focuses on steps farmers can take.

## Weaving Out

Here are three ways that we weave out, extending our farm into the community.

### 1. SELL HYPERLOCALLY.

Many small farms tragically fail to tap into sales outlets right in their backyards, opting instead for ostensibly more profitable markets farther away. But with a bit of research, you might be surprised at the sales opportunities close by.

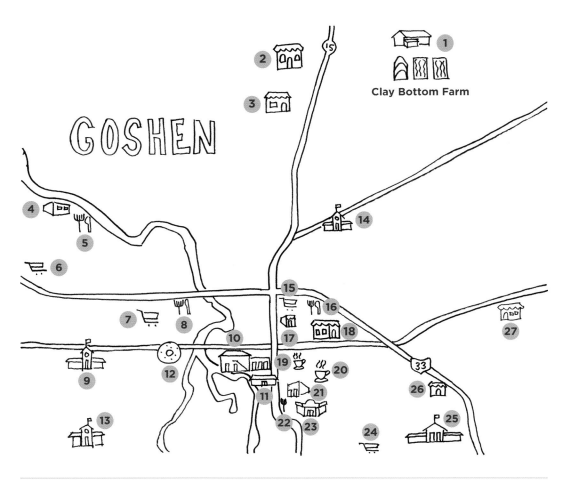

This map of Goshen, Indiana (pop. 40,000), shows just a small selection of potential food buyers within a few miles of our farm. 1. Clay Bottom Farm, 2. Taqueria El Maza, 3. Massimo's Pizzeria, 4. Bread & Chocolate, 5. El Zocalo, 6. Martin's Super Market, 7. Kroger, 8. Castañeda Tacos, 9. West Goshen Elementary School, 10. Goshen Brewing Company, 11. Goshen Farmers Market, 12. Dutch Maid Bakery, 13. Goshen Intermediate School, 14. Chamberlain Elementary School, 15. San Jose Supermarket, 16. Si Señor, 17. Olympia Candy Kitchen, 18. Venturi, 19. Embassy Coffee, 20. Electric Brew, 21. Constant Spring, 22. Los Primos Mexican Grill, 23. Maple City Market, 24. Martin's Super Market, 25. Goshen High School, 26. La Perla Tapatia, 27. Mi Poblanita.

Here is a simple, practical way to geolocate hyperlocal customers. Enter your farm into Google Maps, then search for restaurants, grocery stores, farmers markets, and institutions with cafeterias (like schools and colleges) within a small radius, say 5 miles. Then visit as many as you can. Above is a map of Goshen, Indiana, a small city of 40,000, showing just a sampling of potential food buyers within a few miles of our farm. In fact, there were too many to fit into the map: 12 grocery stores, more than 50 restaurants, 12 schools,

| Clay Bottom Farm Value Sheet | | |
|---|---|---|
| **Chef Name:** Jesse | | |
| **Preferred communication method and time:** texts noon–4 p.m. | | |
| **What** do you want? Be specific. | **When?** Be specific. | **How much?** Be specific. |
| **Bulb fennel,** tops left on, no roots | Tues and Fri noon–4, Sept–Jan | About 10 per week |
| **Salad mix,** any kind; just not too much spinach | Tues and Fri noon–4, all year | 9# per week |
| **Tomatoes**—mix of colors, lots of green zebra, ripe | Tues and Fri noon–4, May–Oct | 50–60# per week May–June 30# per week July–Oct |
| | | |
| | | |
| | | |
| | | |
| | | |
| | | |

We meet with chefs at least once per season. Together, we fill out this value sheet.

2 wedding venues—in addition to a thriving farmers market. These establishments buy and serve food every day. Many are locally owned and more than happy to buy local food. Try creating a similar map for your farm. You might be surprised at the number of customers right in your backyard.

When we moved, Rachel; our assistant farm manager, Nicole; and I revisited each of our accounts, and we reached out to a few new establishments in Goshen. We carried with us the sheet above.

We asked the food buyer at each place: What produce do you want? When do you want it? How much? We sought specific answers. If they said tomatoes, we'd follow up and ask: What size, shape, and color do you prefer? When do you want the food delivered? Tuesday morning? Friday afternoon? How much might you want in each delivery? We sought specific answers to help us plan better.

After a few days of interviewing, we had in hand a stack of value sheets—the start of a precise and profitable farm plan.

We work hard to precisely understand what value means to our customers.

The next step was to winnow the value sheets, selecting the vital customers and the vital crops, using the criteria I discussed in chapter 4. The value sheets aren't binding legal documents. We make it clear to customers that we will do our best to fill their orders, but factors like the weather can always foil our plans.

Now, twice each week, we send out a fresh list to the customers that we chose, collect

**Table 5.1.** Clay Bottom Farm Fresh List

| | Price per Case or Pound | Case Size | Cases Available |
|---|---|---|---|
| **TOMATOES** | | | |
| Heritage tomatoes, 1sts | $45 | 10 pounds | 12 |
| Heritage tomatoes, 2nds | $35 | 10 pounds | 8 |
| Cherry tomatoes | $50 | 10 pounds | 12 |
| Red slicing tomatoes | $30 | 10 pounds | 6 |
| Yellow slicing tomatoes | $30 | 10 pounds | 4 |
| **FRESH GREENS** | | | |
| Spring mix, bulk pounds | $11 | 1 pound | 30 |
| Spring mix, clamshells (5 ounces) | $70 | 18 clamshells | 6 |
| Curly kale, bulk pounds | $8 | 1 pound | 20 |
| Curly kale, bunches | $36 | 15 bunches | 4 |
| **HERBS** | | | |
| Cilantro, bulk pounds | $20 | 1 pound | 8 |
| Italian basil, bulk pounds | $18 | 1 pound | 10 |
| **FRESH ROOTS** | | | |
| Carrots, bulk pounds | $30 | 15 pounds | 4 |
| Hakurei turnips, bunches | $30 | 15 bunches | 6 |
| Radish, bunches | $30 | 15 bunches | 10 |

Sample Clay Bottom Farm fresh list. We text a form like this to local customers twice each week, and we collect their orders on the same spreadsheet.

their orders, and deliver the food they said they wanted, typically within four hours of harvest—faster than an Amazon truck. To assemble orders, we use a spreadsheet on Google Drive, similar to table 5.1.

Of course, communities are always in flux. Restaurants open and close all the time. Each season, we pick new chefs and food buyers to interview, and we shuffle our mix of crops and customers. We meet with each account once annually to review their value sheet: Did a menu item change, is the delivery schedule working, are there other vegetables they'd like us to grow?

This approach to farming decenters the farmer and puts the community out front. It might seem to strip the joy from farming because it puts you, the farmer, in the passenger seat. In some ways, this is true. Sometimes I feel like I work for our chefs because I take so many orders from them. But there is freedom in this approach. I don't need to constantly decide what to grow or how much. More important, this is a way to respect the local community and to weave our farm into it.

## 2. USE LOCAL RESOURCES FIRST.

Diligently searching for local customers shortens the distance that our products have to travel off the farm. But we also want to shorten the distance of things that come onto the farm.

In our move, we wanted to shorten the distance of things that came onto the farm. We compost local leaves for fertility, and we no longer truck in minerals from far away.

## From Independence to Interdependence

In the past, my default vision was for a separate and independent small farm. As a teenager, I imagined what it would be like to create a radically self-reliant homestead where I grew all my own food, built my own house, made my own clothes. Now I realize that we are always connected to people around us. Very few folks—probably lonely hermits—can claim that they never rely on others for food, tools, building supplies, or social support. The question is whether we will intentionally strengthen those connections or disregard them.

*Independence* means unfettered self-reliance, without a sense of dependence on others. *Interdependence* implies that farmers and local communities should interweave. They should rely on each other for mutual gain. Author adrienne maree brown writes, "Most of us are socialized toward *in*dependence—pulling ourselves up by our bootstraps, working on our own to develop, to survive, to win at life. . . . The idea of interdependence is that we can meet each other's needs in a variety of ways, that we can truly lean on others and they can lean on us."[9] Mindfulness teacher Thich Nhat Hanh created the word *inter-be* to express the concept: "You cannot just be by yourself alone," he wrote. "You have to inter-be with every other thing."[10]

In fact, Nhat Hanh goes so far as to explain that independence is an illusory impossibility. Just as waves cannot be truly separate from the ocean, a person cannot ever be truly disconnected from other people. This is a foundational Buddhist insight.

There have always existed social models of interdependence alongside the ideal of independence. For example, the Pokagon Band of the Potawatomi people in the upper Midwest organized their work according to clans, or groups of families. Each clan had a role to play in the community. For example, bear clan managed security, mediation, and collecting medicines for healing.[11] Food production in the Pokagon tribe happened within the context of clans, and food producers were thus an integrated and visible part of the community. The tribe is currently working to revitalize the clan structure.

In the Mennonite farm community where I grew up, economic incentives pressured families to farm alone and get big, but the community also valued interdependence. If a farmer's barn burned down, the community joined together to rebuild. When a greenhouse that we built blew on top of our barn roof, our Amish neighbors arrived within minutes to help us take it down. As I started writing this book, my dad entered the hospital for surgery; when I stopped in to check on him recently, I found one of his neighbors mowing the yard while another neighbor stopped in to ask if there was anything he could do.

Maybe you live in a place where the neighbors might not stop by. What can one person do? You can "start with hello," as Shannon Martin, an author from Goshen, writes.[12] You can get to know the people who live next to you as a first step.

The work of localizing really amounts to noticing and strengthening ties with the people who live nearest to you. Interdependence is stronger than independence. And in the long run, it is a more satisfying way to farm.

When we moved, we were still using kelp from the Atlantic Ocean, fish fertilizers from Maine, and minerals mined from even farther away. If we want to reduce our food miles, shouldn't we be concerned about the miles these products travel as well? We were emboldened, reading *Small Is Beautiful*, to localize fertility. While we were already using local compost as part of our fertility plan, we decided at the new farm to phase out other fertility sources. We now use our own compost exclusively for fertility in the field. Fortunately, the city of Goshen is willing to deliver leaves for composting, collected from within a mile of our farm, for free. A local brewing company, which is also a customer, delivers to us spent grains that also go into our compost piles.

Our work in the years ahead is to replace other inputs with local alternatives. Could we replace our seed starting mix, from Vermont Compost Company, with our own? Could we replace row covers, which we use to protect crops from frost, with local reed or hemp fiber mats? Could we replace plastic drip tapes, which we use to irrigate, with an Egyptian-like canal system? Could we source more of our seeds from our community or from our own farm? Could a local blacksmith forge our next generation of tools? Could we use local wood for tomato stakes, local willow to make delivery baskets? This is the "opposite direction" that Schumacher envisions, and farming like this will entail solutions that we haven't yet envisioned.

Before, when we farmed on more land, ideas like these might have seemed far-fetched. But as we shrink our production footprint, these ideas seem increasingly plausible. Our fields, before our move, were just too big for creative solutions. And the idea of a locally produced hemp mat sounded like a granola-hippie dream (though we do love granola and try to be hip). At a micro scale, a truly localized farm seems more doable.

## 3. PARTNER TO MAKE FOOD AVAILABLE TO LOW-INCOME FOLKS AND KIDS.

Weaving out should be expansive and include members of the community who couldn't normally afford organic food. At our farmers market, vendors and local nonprofits cooperate to triple the money spent by customers using SNAP vouchers, up to $20—so a $20 SNAP voucher is worth $60 when spent on local food. Patients at a local community health clinic also receive free $15 weekly vouchers to spend at the farmers market.

The market also sponsors a Sprouts Kids Club, where kids receive two free $1 tokens per week to spend on fresh fruits and vegetables. Why is it good for kids to get their own tokens and choose their food? We've seen

Visiting students from Teen Growers, a local gardening internship program for teens, try out tools at Clay Bottom Farm. Providing education is a way to integrate into the community.

seven-year-olds *choose* kale when they have tokens; something they might otherwise complain about becomes a new thing to try. These programs show that farmers and communities, woven together, can make a real impact by making fresh vegetables accessible to children and people with low incomes.

## Weaving In

Weaving in means opening your farm to folks in the surrounding community. I've found that bringing others onto the property enlivens the farm. Visitors give us ideas for how we might improve our growing, and their delight—especially from kids—reminds us why we farm. Here are four ways that we weave in.

### 1. EDUCATE.

By educating, a micro farm can have a big reach. We host school groups on our farm and keep the tours lively by engaging students in hands-on projects. We use a mutual learning model, where we encourage sharing and conversations, sometimes blurring the line between teacher and student. My favorite way to lead groups is to start by asking folks to share about their family foodways: Did your ancestors farm or grow food? What are their stories? What food traditions did you grow up with? I am always impressed with the diversity of answers.

For people interested in starting new market gardens, we host two-day Lean Farm Start-Up workshops, intensive training designed to help new growers be profitable right away. While participants come from around the world, we also save spots for local folks to join. Again, these are mutual learning experiences. I often share problems we might have in our production and solicit ideas from the group, thus improving our farm.

Of course, having people on the farm takes time and focus away from production. It's important to set boundaries and to ask for compensation

First and second graders from Bethany Christian Schools harvest tomatoes at Clay Bottom Farm.

when appropriate. We rarely host groups during our peak season of June through October, and we have learned to politely say no when the timing isn't right or when we simply need a break from visitors.

## 2. PROVIDE GOOD JOBS.

Employees, even if they are part-time, provide a strong link to the local community. We employ two part-time workers year-round on our farm. We pay living wages and send them home with fresh vegetables and eggs. We write clear job descriptions, which we review with workers annually.

Here are practices that have worked well for us when hiring and managing workers in a micro farm context:

### Hire at the right time: after you have steady work.

The best time to take the plunge and hire is when you have enough value-adding, standardized work to keep employees productive. The beginning

A meal with visitors at Clay Bottom Farm. A farm should be shared and enjoyed with the community.

stages of a farm involve a lot of systems design, experimentation, and mistakes. This can be a chaotic environment in which to bring on paid staff. Refine your systems first, then hire.

### Hire for specific tasks.

We don't usually hire general farmworkers but rather people who are suited to help in a specific way. For example, last winter we hired a college student with construction skills to help me build our second greenhouse. The student needed work for one semester, about the length of time needed for the project. The student's work was exclusively focused on the greenhouse, not on seeding lettuce or delivering tomatoes. We've also hired short-term workers to help with website work, video editing, photography, and designing our product labels.

We hire a college student each summer to complete five specific tasks: washing greens, trellising tomatoes, harvesting tomatoes, delivering food to restaurants, and washing dirty totes. Nicole Craig, our assistant farm manager, helps us perform management-level work: organizing workshops and our CSA, communicating with customers, and collecting payments.

Be flexible with hours as much as possible.

Our entire crew assembles at the farm Tuesday and Thursday mornings to harvest and deliver food. But most other farmwork is performed at flexible times. For example, trellising tomatoes doesn't need to happen at precisely 8 a.m. each Wednesday; for this task, we allow workers to complete the job when it suits, so long as it happens once each week. Other flex-time examples include spreading compost and mowing walkways. Nicole performs much of her management work from home, where she is homeschooling two kids, because many of these tasks can happen in a wide window of time from anywhere. It is easier to hire part-time workers and retain them if you can be flexible with hours.

## 3. HOST VOLUNTEERS AND INTERNS.

Volunteer and internship opportunities can fill a community need. We use paid employment for most farmworkers, but we still accept a student volunteer each semester through a program that pairs high school seniors with volunteer work placements in areas of their interest. Several of these students have gone on to study agriculture in college. We also partner with a local mental health organization as a volunteer placement site for persons struggling from mental illness. And our farm hosts young people from around the world through a Mennonite development organization. These partnerships are mutually beneficial and are another way to weave into the community beyond the farm.

## 4. TEACH IN PRISONS.

People who are incarcerated are part of the community, too. There are a lot of ways to include prisons and jails in your work; a good starting point is to talk with jail chaplains and ask if there's a program you can join.

We've partnered with University of Notre Dame's Moreau College Initiative at Westville Correctional Facility in Westville, Indiana. Through the program, students inside the prison can earn college degrees. Professors teach classes in many subjects, including sustainable agriculture. In fact, students enrolled in the program have organized a prison garden. The

Westville garden is large, productive, and a place for folks to unwind and to learn skills they can take with them when they leave.

At the invitation of students and professors, I've presented at Westville about our farm and answered questions about growing crops and running a small farm business. A student in the Moreau College Initiative, Shannon McDevitt, helped transcribe an interview for this book. We have also hired folks recently released from prison on our farm. Often, it's difficult for these folks to find work, and after a season on the farm, I can provide a reference for future employment.

Schumacher's vision for an "economics as if people mattered"—inspired by the Indian movement for independence and rooted in an ethic of building up local communities—inspired us to integrate more deeply into Goshen. In the ways I described here, we've tried to localize to the extent that we can. Our goal is to train our vision more and more each season on achievable things that we can do close to home, with others who live nearby, things that increase the self-sufficiency of our farm and the self-reliance of our village. This is Swadeshi.

# A Tiny Farm in Chile

*"Give them a taste of what you have versus what they're getting at the market, and it's like a whole world opens up."*

Tamara Bogolasky owns El Borde, a market garden about three hours from Santiago, Chile, where she grew up. Like Tiny Giant Farm, her growing area is a yard-size ⅓ acre.

When she was young, however, farming wasn't on Bogolasky's mind. "Santiago was a big city, and food was just always available," she told me. She studied advertising in college at the Universidad del Pacífico. At age 22, just after delivering her thesis, she boarded a plane for New York City. Her goal was to be a photographer.

While she trained in photography, she also dabbled in growing food. She grew herbs in the windowsill of her Brooklyn apartment; in 2014, she helped start a permaculture garden—Imani Gardens in Greenpoint, Brooklyn—and she attended Farm School NYC. Growing food "felt easy," she said. It surprised her.

She wondered about changing careers. "My work in photography and the arts wasn't going where I wanted it to," she said. Then one day, "the flash of my future came through my head." She had a vision of gardens, with tourist lodges—"a whole project"—and she realized, in that moment, that she was in the wrong place. Implementing that dream made more sense in Chile than in New York. In 2017, she boarded a plane back home.

## LOCALIZING CHILE

Chile is one of the longest and skinniest countries in the world. It spans 2,670 miles north to south and averages just 110 miles east to west. It covers three continents, claiming territory in South America, Antarctica, and Oceania. Geographically, it's a wild ride, as diverse as it gets. Inside Chile, you'll find the driest nonpolar desert in the world; a central valley teaming with vineyards, orchards full of peach, plum, and apricot trees, and giant fields of honeydew melons; pristine lakes; thick rainforests and lagoons; the majestic Patagonia, with vast grasslands, deserts, and glacial fjords; and the Andes Mountains—not to mention 91 volcanos and 20,000 square kilometers of icebergs that sprawl into the sea.

Chile's food scene, however, is anything but diverse, according to Bogolasky. For decades, it's been dominated by monocrop farms, imported food, and a lack of food choices, despite the Valle Central's optimum growing conditions and many farms.

The gardens at El Borde, in O'Higgins, Chile. Photo courtesy of Tamara Bogolasky.

But that's changing. Bogolasky is part of a new generation transforming Chile's food culture. Bogolasky, with a group of Chilean market gardeners, visited Clay Bottom Farm in 2022 on a learning tour. Afterward, the group invited me to their WhatsApp group, where the farmers swap photos of their produce and share advice about combatting pests and diseases. They buy and sell greenhouses from each other. They banter. One farm—Huerto Cuatro Estaciones (Four Seasons Farm)—hosts workshops for new growers and has built a Spanish-language market-gardening online course. The passion for change is palpable.

Bogolasky bought land in O'Higgins, Chile, in a suburb of the village of Santa Anna, in the Central Valley, surrounded by vineyards. She was excited to finally start her farm, but there was a problem: People in the Central Valley weren't familiar with her food. "Seven years ago, nobody here knew that kale existed," she said.

Bogolasky now sells most of her food in Santiago, where it was easier, in the beginning, to find customers willing to try new vegetables. But her goal is to tighten her delivery radius, to sell more food every season closer and closer to home. By now, 25 percent of her food stays within an hour of the farm—she sells in Pichilemu, the surfing destination, and in Santa Cruz.

Bogolasky told me she uses three strategies to localize. First, she gives food to potential customers to try. "People here have habits that are very hard to break," she said. They've become accustomed to trucked-in food. "But give them a taste of what you have versus what they're getting at the market, and it's like a whole world opens up."

Tamara Bogolasky, owner and operator of El Borde. Photo courtesy of Tamara Bogolasky.

Second, she makes it easy for customers to get her food. In Chile, she said, "A lot of people are stressed. They work long hours to meet their basic needs. If you ask people to go pick up your food, then you lose a lot of people." So, she takes the food to them. Eighty percent of her sales happen through a direct-delivery CSA, where she or a hired hand drops food directly onto the customer's porch.

Third, she opens her farm to school groups, and she is working with the local municipality to create space on the farm for community members to grow food. She is actively working on an on-farm tourist venture where locals can see that it's possible to grow food organically.

On her farm's WhatsApp group, customers trade recipes and can see photos, which Bogolasky shares, of their food in the field where it's grown. "They know that I am committed to them, and they are committed to me," she said. "We've formed a community that sticks with it for the season."

For Bogolasky, changing the food system is a group effort of her customers, her municipality, local schools, and other farmers:

> *This style of farming is really new. El Borde is only six years old, and a lot of farms have started since we began. Chile is a beautiful place, and we're all learning from each other how to make it better.*

# A Romanian Plum Farmer

*"We've been farming here in Transylvania
since the beginning of time."*

Attila Szocs is an avid plum farmer from the Transylvania region of Romania. He and his wife, Erika, integrate their farm into the local village almost by default—for centuries, farmers in the region have banded together.

Szocs is also president of Eco Ruralis, an organization of 19,000 Romanian food producers that advocates for peasant farmers, and a leader at La Vía Campesina, an international small-farm advocacy group.

Szocs traces his farming roots to the 1600s. His ancestors farmed under a ruler before becoming "free peasants." "My ancestors always had community-oriented subsistence farms," he told me. "They were producing vegetables, fruits, animals—everything—like millions of others."

While farms operated independently, they were also deeply interdependent. Farmers relied on each other to share tools, land, draft animals, and seeds. They sold their wares, grains, fruits, and vegetables to each other at markets in the center of the village. "This was like an exchange," Szocs said. "And this went on quite unnoticeably until the Communist era, when everything changed."

The Communists took away land, including from Szocs's grandparents on both sides of the family. His mother's father refused to leave the land and was beaten and eventually forced off his farm. He became depressed and, in an act of protest, committed suicide.

"His story pushed me into farming, to reconnect with the land," Szocs told me. "I was inspired by the idea that you can be so connected to a piece of land that you become willing to sacrifice your life for it."

In 2007, he and Erika bought a ½-hectare farm near his ancestral homeland. They tend an orchard with 75 plum trees, their main source of income.

Erika Andreea Szocs-Boruss preparing a batch of plums for *magiun*, a traditional Romanian plum jam. Photo courtesy of Attila Szocs.

They are largely self-sufficient in what they eat, maintaining a large garden of tomatoes, potatoes, cabbage, carrots, strawberries, and more.

## INTEGRATED FARM AND VILLAGE

The farm is deeply woven into the village to the point where village and farm are nearly inseparable.

Attila and Erika share land with other villagers. They've joined a community trust, which gives them access to 50 hectares, in addition to their farm, for growing food and raising animals, a venture they plan for the future. They also share land with a neighbor to grow potatoes.

They share seeds and tree seedlings as well. Szocs grows vegetable seeds that his grandmother saved through the Communist era, and he and other village farmers have started a seed bank of heritage seeds from the region. He is collaborating with his neighbors on a separate project to save traditional plum varieties from Transylvania.

Attila and Erika share use of a community distillery, which dates to the 1700s, to process plums. "It's the same distillery, just patched up to

A Romanian plum orchard. Photo by Getty Images/xalanx.

104

work for generations," he said. "I am using infrastructure that was set up for me by my ancestors. While I reinvent little things or innovate little things, my cultural bearing, my farming, is the way of the region. It's a spiritual thing."

Each fall, he and Erika collect their plums into about 50 barrels, which they ferment for two or three months and then turn into *palincă*, or plum brandy. Scozs emphasized to me that he could not do it alone. To distill the plums, a *palincar*, or master distiller—a descendent of the family that originally built the distillery—tastes the brandy at each step. In fact, no one but the *palincar* is allowed to taste the brandy until it is finished.

> We all use the same palincar *because he tastes brandy all day long, from all of our farms. He knows when to separate the rezes [foreshots] from the* rachiu *[feints] from the* palincă, *so the taste is pure. He has done this work for more than 20 years, and before that he sat watching his father for 20 years, and his father sat watching his grandfather, and on and on. Unfortunately, they all die at about age 65 of liver problems. This a hazard of the job.*

I got the sense, listening to him, that brandy, for a good many Transylvanian farmers, is more than an income—it is heritage in a bottle. Listen to Szocs describe a sip of good Transylvania *palincă*:

> To drink our palincă, *you put it in a small cup and then you put a bit on the tip of your tongue, but you don't savor it like whiskey. You throw it to the back of your throat and swallow it, with one shot. On the way down it should burn you a bit, but not in a way that makes you cough. It should burn you in a way that heats you up from the inside. It's like a sun burning inside of you.*

He added:

> We've been farming here in Transylvania since the beginning of time. Even now, 7 million out of 19 million people in Romania are engaged in agriculture. We have huge cultural traditions in music, dancing, and of course food and drink, and rural people keep these alive. We are the ones who give them to the next generation.

## UKRAINE AND SWADESHI

Szocs works closely with Ukrainian small farmers in his nonprofit role. Szocs told me that when Russia invaded Ukraine in 2022, "the big farmers left, and the small farmers stayed behind to feed the people." Over half of Ukraine's 14.7 million households are involved in small-scale agriculture, and these farmers produce almost all of the vegetables eaten in Ukraine. Most meat and milk are also produced on small family farms distributed throughout the country.

When Szocs asked his Ukrainian friends what they needed, they told him, "Seed, because many of the seed stocks were damaged in the war." So, from their village seed banks, Romanian farmers are supplying potato and heirloom vegetable seeds, and they are paying for gas to transport food from one Ukrainian village to another and from villages to cities.

To Szocs, a healthy self-sustaining, self-reliant village doesn't just care for itself. Bounty should spill over. Part of self-reliance is considering needs outside of yourself. He said, "My grandfather always told me, 'When you farm, you tend to work with your head down, looking at the ground. Now it's time to farm with your head up.'"

# CHAPTER 6

# No-Till for Micro Farms: The Deep-Mulch Method

*A* *note about part 2:* In the remaining chapters, I'll explain in detail how we applied the principles from part 1 to set up our new ⅓-acre farm—for readers looking to start a micro farm and for serious home gardeners looking for ideas to improve their growing. Schumacher repeatedly said: "First must come the work—intellectual effort. Then the word must become flesh."[1] This is the flesh.

I'm not intending to offer a comprehensive guide for market gardening. Rather, these are the absolute essentials. Here's what you really need to know to get started.

In chapter 10, I offer a plan for selling $20,000 from your backyard, with specific crop-growing advice. This chapter takes the methods in this book and condenses them to an even smaller, more approachable scale.

▪ ▪ ▪

In part 1, I argued that a key to getting small is to simplify—to find the most straightforward solution to any problem. The biggest change we made in our production when we shrank our farm was to simplify our growing using the deep-mulch method.

Here's how it works: To start new plots, lay down 4″ of good compost right on the surface (assuming the ground is already worked up). Don't till it in. Just rake it smooth and grow right into it. Every two years, or as your crops indicate the need, add another 1″ of compost.

Simple. That's it. This is the best way I know of to grow high-value crops at a low cost on a small plot of land. It is a straightforward solution—powerfully effective but with few steps.

With deep mulching, you suppress weeds, create friable (loose) soil, and give plants a source of minerals for years. The method harnesses the

Rachel harvesting greens. Greens in this picture are grown on 4" of compost, made from local leaves that we placed and left directly on the surface.

power of biology in the soil to rejuvenate the rhizosphere, the zone of soil around roots—it is a way to farm like a tree. The only tools required are a shovel and a rake. The method replaces trucked-in fertilizers like kelp, fish emulsion, and calcium with local leaves. It passes the Swadeshi test.

Below, I'll show how to farm with the deep-mulch method. In the next chapter, I'll show the two-step bed flipping approach that complements deep mulching. Both are no-till practices that incorporate new understandings of soil science to grow better crops with less work.

# Building Plots with 4" of Compost

Building plots with deep levels of compost costs more and takes more time initially, but the benefits pay off. I'll discuss this approach assuming 50'-long beds, but the principles are simple and can also be applied to 4' × 8' raised beds. Applying deep compost only makes sense on tiny farms—it is impractical to cover large farms with so much bulk material.

## STEP 1: WORK UP EXISTING GROUND.

If your garden is already worked up, then skip ahead to step 2. Otherwise, you will need to terminate existing vegetation and loosen the ground first.

Step 1 on our farm was to mark out plots and loosen the ground with a tiller.
Photo courtesy of Adam Derstine.

In most cases, the best approach is to mow vegetation as low as possible and then till it. This first, and maybe only, tilling will clean the field and create an initially loose soil structure. Three or more passes with the tiller might be required to ensure the vegetation won't grow back. But to be sure, wait two weeks after tilling to see if any grasses sprout. Don't skimp here. You don't want to cover living grasses with compost—the grass will find a way through.

If the land you need to loosen is a perennial pasture, then plowing might be required first. Plowing will rip up the deep and tenacious roots found in these pastures more effectively than tilling. After the sod has been flipped, wait until it has dried—typically within a week—and then till or disk until the ground is smooth. Both tilling and plowing are tasks that you can hire out to a landscaper if you don't have access to the equipment.

Another option: Terminate existing vegetation with black silage tarps. Terminating with tarps can take a few weeks or a few months depending on the weather. Don't hurry—grass must be completely burned down. Silage tarps are thick plastic sheets used for multiple purposes in agriculture. They are UV-resistant and come in a variety of sizes. For terminating vegetation, use a size that covers your entire plot, if possible. The tarps kill vegetation by depriving it of light and by cooking it. Make sure the edges are pinned down well; for this purpose, we use sandbags placed about every 6′ around the edge. If the ground is firm after you've removed the tarps, then I recommend using broadforks, digging forks, or a tiller to loosen it.

On very small patches—say 1,000 square feet or less—existing vegetation can be terminated by laying down cardboard covered with 4″ of compost. This is more cumbersome than tarping or tilling, and I don't recommend it unless you have a handy source of cardboard. Also, compacted ground underneath the cardboard might be impossible for some plant roots to penetrate.

## STEP 2: ADD THE COMPOST.

To add the initial 4″ of compost, even on tiny plots, I recommend the use of a skid loader or a tractor with a bucket. Alternatively, if you are buying compost, ask the driver to dump—but not drive—directly onto your plot.

Next, grab a sturdy aluminum scoop shovel or compost fork and spread the compost as evenly as you can. We've found that this is a fun group project—perfect for visiting school-age kids—because it is rewarding to see the compost build up. Simple tasks like this are opportunities to weave the local community into your farm.

An option is to use wheelbarrows to move compost around the plot. However, we don't like the compaction that wheelbarrows leave behind. On a tiny farm, less than 1 acre, spreading by hand is manageable.

We never drive heavy equipment on our plots because the equipment will compact the ground, making it difficult for roots to penetrate. If the ground is frozen hard, however, equipment weight doesn't matter. In fact,

We use the deep-mulch method in greenhouses and in the field. Here we are spreading a deep layer of compost in a greenhouse.

we sometimes add compost in the middle of winter because we can drive all over our plots with our skid loader.

If you want to spread deep mulch over an acre or more, you'll need a compost spreader. Some growers use manure spreaders. Community Machinery in Sackville, New Brunswick, manufactures compost spreaders designed by Broadfork Farm in Nova Scotia. We sometimes use a compost spreader attachment on our skid loader. The implement is made by Ideal Welding in Middlebury, Indiana. A flap in the bottom of the bucket opens up, letting compost fall out in small amounts. An auger inside the bucket stirs and breaks up clumpy compost, making it easier to slip through the opening. These machines can be found online.

Adding compost in the winter, when the ground is frozen, will not compact the soil. The compost stays unfrozen due to its internal heat.

Never till in the compost. Think like a tree: Leave organic matter on the surface and let it decompose. If you till the compost in, you weaken its natural structure and you move valuable minerals from the compost to below a depth where plant roots can reach them. Also, tilling brings weed seeds up to the surface. We have performed side-by-side tests of beds with compost tilled in and compost left on the surface. Crops consistently performed better when we left the compost alone.

For this step, go ahead and cover the whole plot evenly. Don't worry about paths just yet—that will come later.

### How to determine how much compost you need

To determine how much compost to move onto your plot, I recommend using the calculator created by Green Mountain Compost, found at www.greenmountaincompost.com/compost-calculator. Per 1,000 square feet, you will need 12.4 cubic yards of compost to achieve a depth of 4″. For our 15,000 square foot plot, we used about 200 yards of compost.

### Why 4″?

Because compost at 4″ deep is capable of smothering most weeds. Also, 4″ compost can be worked without bringing native soil up because most hoes

HOME GARDENER MODIFICATION
## How Home Gardeners Can Use the Deep-Mulch Method

The deep-mulch method is the best system to use in a home garden. The hurdle to overcome, for many, will be making the decision to invest in enough compost to cover the entire garden by 4″, but it is worth it. Remember that with deep compost, you will never again need to invest in fertilizers and amendments. The deep mulch will retain water well, dramatically reducing the need to irrigate. Also, yields will immediately increase. All these factors make the investment easily worthwhile.

I recommend asking the farmers at your local farmers market where to source good compost. Really, the compost you use does not need to be perfect—it can be made out of many types of material (though avoid straight manure), and

chunks are fine, as I explain on page 121. Ideally, you can have the compost dropped right onto your garden.

Spread the compost evenly over the whole garden, using a wheelbarrow and shovel. There is no need to create raised beds unless you want raised growing areas to reduce bending over or to improve drainage in a low area. You can cover pathways with cardboard and wood chips, if you prefer the look, or leave them in compost.

What if all you have is a postage stamp of a yard? Build a deep-compost garden bed! Simply lay down a large piece of cardboard and build a frame from 2″ × 6″ lumber on top of it. Then fill the frame with compost. That's it. You can plant straight into the compost.

and other field tools won't penetrate below 4″. Most importantly, this depth creates an ideal soil structure for most plants.

### What kind of compost should I use?

When buying compost, localize. Many municipalities now make and sell compost. If you are making your own (see page 119 for instructions), use materials sourced close to home. Use mostly vegetative carbon (C), or "brown," material, like leaves and straw, with little or no animal manures (or other "green" material), because too much animal manure can overload soil with nitrogen and phosphorus. Also, compost with a high carbon-to-nitrogen ratio will have a higher amount of fungal activity. We stick with the leaf-based compost because, as I said earlier, the leaves are local and free.

In our experience, for optimum plant growth, compost should be used within one year and about 80 percent finished; at this stage, the compost is

## How Deep Mulching Improves Soil Chemistry

Deep mulching improves soil chemistry because good compost is high in nutrients—like magnesium, potassium, and calcium—with a balanced pH (between 5.5 and 8).[2] Before we added our deep mulch, our soil tests showed bleak results: pH was off, nutrient levels were low, cation exchange capacity (CEC)—a measure of the soil's ability to absorb nutrients—was abysmal. After adding 4″ of compost, nutrient levels and CEC registered above average and the soil was pH-balanced.

## How Deep Mulching Improves the Physical Structure of Soil

Deep mulching improves structure by adding fluffy material to the soil surface. You essentially bypass problems associated with sandy or clayey soil. Soils retain an optimum structure for growing plants for many years as long as you don't till and break apart the fungal networks that hold it in place. Tilling compost is like cutting framing members out of a house— the house will come crashing down.

On land that is very clayey or very sandy, I recommend applying 6″ or even 8″ of compost. Deeper levels of compost can also improve drainage in low areas. I have never seen benefits beyond 8″ deep. In most cases, 4″ as a minimum is the right amount.

On some beds, we added as much as 8″ of compost to improve structure and drainage.

at its peak, ready to unleash minerals for many seasons. If you buy old compost, you are just paying for heavy bulk material because its minerals will have likely leached away.

Compost for gardening does not need to be "potting mix" perfect. A bit chunky is fine as long as you can run a seeder through it. Chunky compost will finish breaking down within a season of application. Don't burden

Deep compost on a garden plot. Compost for deep mulching does not need to be "potting mix" perfect. Chunks and small sticks will break down over time.

yourself. Trees don't pulverize their dead leaves before using them; you don't need to, either.

Do avoid compost that is matted and slimy. This indicates it was poorly made, without sufficient turns. There is a difference between a rot pile and a compost pile.

Important tip: Compost for deep mulching must have zero weeds. Weedy compost will bog down your farm, potentially for years. Test first. If possible, bring a small load to your farm and grow a bed of crops in it. Did weeds grow out of the compost? Do plants appear healthy? If you don't have time for this field testing, then fill a potting tray with compost and set it in a south window or in a germination chamber for two weeks, watching for weed germination. The

Lettuce seeds germinating in deep-mulch beds. Don't till compost in; leave it on the surface and grow right into it.

composting facility should be able to tell you if there are weeds in the mix, but we have learned the hard way not to always trust what we are told.

## A doable, minimalist method for making compost

If good compost is not available for sale in your area, don't despair. Compost for deep mulching can be easily made on-farm as long as you have access to raw materials and a skid loader or tractor with a bucket, and it is typically ready within one year.

We use a version of Steven Wisbaum's "low-input composting system" to make our compost. Wisbaum is a professional composter with decades of experience; you can read more about his approach online.[3] The low-input composting system takes a *simple is best* approach, and the compost it produces is higher quality than any we can buy locally.

In brief, the system relies on a "small number of well-timed, thorough turns." No fancy equipment is required other than a tractor with a bucket or a skid loader. Again, this is a way to farm with few tools. We turn our piles just three times before they are applied to the field. Here are the two steps that we use.

### 1. Collect raw materials.

To find low-cost composting ingredients, start by calling your local street department or local landscapers. Municipal street departments can tell you where to find leaves. Spent grains from breweries can often be had for free.

Other low-cost ingredient ideas: grass clippings, poor-quality straw, moldy hay bale "seconds," and leftover produce from grocery stores. Sometimes entire fields of hay are cut and discarded if the hay is too wet to bale and store. This is an excellent way to turn waste into a useful product. I've found that local truckers, who haul bulk material on a daily basis, are the best resource for finding low-cost compost-making ingredients.

As a general rule, bulk material will decompose to about one-fourth its mass. If you require 100 yards of compost, collect 400 yards of raw material. However, because raw materials come in a variety of weights and shapes, shrinkage can vary widely.

It's best to build one large heap rather than several small ones. We build one heap each year. Our heaps

We arranged for the city to dump leaves for free on our property.

are about 9' wide, 6' tall, and 50' long. The limiting factor is your equipment: Build piles as tall as your equipment will allow. The reason for tall piles is to create a large core—the center of the pile where the composting action takes place.

*2. Turn three times.*

According to Wisbaum, piles should be turned for four reasons:

a. To mix drier, less-digested outer material with wetter, more-digested inner material and ensure all ingredients spend time in the core, the biologically active area within the pile.
b. To restore pile porosity, that is, to loosen it, which introduces oxygen and increases microbial activity.
c. To break up larger clumps.
d. To even out moisture in the pile.

Timing turns is critical. In our case, we collect leaves in the fall and turn them the next summer—once each in June, July, and August—because summer is when the outer material in the pile tends to dry out. Turning in

We keep weeds from creeping into the compost by skim tilling around the pile.

summer circulates material from the matted inner core throughout the pile, keeping the piles evenly moist.

We've found that with this method, our leaf piles are sufficiently broken down within about one year after the leaves were delivered. This compost is a bit chunky compared to commercial compost, but we can still direct seed and transplant into it, after it is spread. By spring of the following year, compost in the field is broken down completely and has transformed into rich, dark garden soil.

Another reason to turn compost is to heat it up. Sustained temperatures above 131°F (55°C) will kill weed seeds and pathogens. We don't monitor pile temperatures because our leaf piles are weed-free to begin with. However, if you are trying to turn weedy grasses into compost, make sure the piles heat up to 131°F for three days or more. The best way to heat a pile, besides turning, is to add fresh manure and/or water if the heap is dry.

A key advantage with the low-input method, beyond time savings, is that fungal networks

## HOME GARDENER MODIFICATION
### A Minimalist Method for Making Home Garden Compost

To make compost for a home garden, I recommend this approach. As the season goes on, collect your garden scraps into a pile—4′ × 4′ is a good size. Cut up woodier parts—like kale stems—with a machete as they go into the pile. In the fall, collect all the leaves you can and add them to the pile. The leaves add "brown" matter, or carbon (the vegetable scraps are mostly "green"). Ideally, create alternating layers of browns and greens, like a cake; this helps the pile heat up. You don't need to build sides out of wood or other material, unless you want to for aesthetics.

Next, turn the pile with a digging fork. One turn per year, with a small pile, is usually enough. The compost should be ready in 9 to 12 months. Since you are constantly adding scraps, I suggest creating at least two piles so that one "finished" pile is always ready and waiting. The compost can be rough when you apply it. Chunks, small twigs, and leafy bits are fine, so long as you can plant into it. It'll continue to break down in the garden.

within the piles remain more intact. This more hands-off approach lets nature do the work. As Wisbaum writes:

> *Scientists and farmers alike are increasingly appreciating the role that fungi play in maintaining the health of the soil/plant ecosystem. Therefore, since turning destroys hyphae, which*

*are branching filamentous structures that serve as the main mode of vegetative growth for fungi, compost piles that are turned less frequently will have higher fungal populations compared to piles that are turned more frequently.*[4]

This is a case where a slower, simpler approach achieves more.

### Why we stopped using cover crops

We stopped growing cover crops because, with the deep-mulch method, our soils have all the fertility they need. Also, our beds are full of crops from March until October, if not later—there is no room for cover cropping. As the next chapter explains, we leave old crops—roots and all—in place as much as is practical; our decomposed crops feed the soil and help to build soil structure, as a cover crop would. In this way, less work equals more fertility.

## Skim-Coating Every Two Years

We skim-coat our deep-mulch beds with 1″ of compost every two years, or more frequently if crops indicate the need. Fall is an ideal time for skim-coating. Adding fertility in the fall is another way to farm like a tree—it is a way to work ahead.

A fall application gives the compost a few months to break down further before being planted into, and the compost helps stabilize soils through the winter, acting as a sort of winter jacket. That said, with our tight crop rotations, we don't always have time in the fall to top-dress all beds, so we add compost in other seasons, but we do as many in the fall as we can.

This top-dressing supplies fresh food (carbon) for soil microbes, keeping microbial activity high. In our experience, it is not necessary to apply more compost than this, as long as you've started with a thick base of mulch and as long as you compost your spent crops in place, a practice I'll describe in the next chapter.

For this skim coating, we use the same compost that we use to build beds. We dump a bucket at each end of a bed and then spread the compost evenly across the surface with a shovel. Sometimes we fill a garden cart with compost, remove the back, and let the compost drizzle out. The cart straddles our beds, so there is no compaction. In winter, if the ground is frozen hard, we will spread compost across the surface with the Ideal Welding bucket attachment, straddling beds with the skid loader. We skim-coat beds in the greenhouses by hand or with the aid of the garden cart.

We use a garden cart with the back removed to dump compost. Then we spread the compost to a depth of about 1″ with a bed rake.

This Ideal Welding skid loader attachment features an auger to spread compost. A flap at the bottom of the bucket opens up. This attachment only works with relatively dry compost.

I want to emphasize this point: Compost replaces the need for shipped-in fertilizer. You don't need minerals like gypsum and calcium trucked from far away or costly sprays like fish emulsion. Composting done right can be a complete fertility management system.

## Laying Out Plots

For decades, market gardeners commonly grew crops in long rows because this facilitated the use of mechanical cultivating equipment. In 1989, Eliot Coleman's *New Organic Grower* advocated instead for using 12"-wide pathways separated by 30"-, 60"-, or 120"-wide growing areas. This growing-bed system quickly took off as the modus operandi on small-scale market farms because it allowed for a wider variety of crops to be grown and rotated in and out of a small space. It also produced less compaction because foot traffic was limited to pathways.

A *plot* is simply a group of beds lumped together. Growing in beds—creating uniform growing areas the same width and length for repeated use—is an excellent way to practice the principle of "standardizing work." Here are eight tips to simplify plot building and maintenance.

124

## What's in a Teaspoon of Soil

As of this writing, scientists estimate that a single teaspoon of soil contains 50 billion microbes (50 billion!), including 10,000 different species of organisms. Incredibly, 1 acre of soil can contain a *ton* of active bacteria. Deep mulching keeps these life-forms optimally active.

The interactions between these microbes and plants are complex and still mostly mysterious. However, especially in the past 20 years, soil scientists have begun to map out these interactions in more detail, deepening our knowledge about soil biota. It is outside the scope of this book to explain this soil science in depth, but for good summaries of this new research, I recommend Jen Aron's articles in *Growing for Market* magazine (growingformarket.com); Jesse Frost's useful guide, *The Living Soil Handbook*; and Dr. Elaine Ingham's foundational work at soilfoodweb.com.

### 1. Build plots no wider than 100'.

Plots should be designed for ease of access. You want to quickly get to your beds to tend and harvest crops. Our plots are never wider than 100'—about 15 beds lumped together—so we can quickly encircle a plot on foot.

### 2. Choose a bed width that you can straddle or easily hop across.

A standard bed width on market farms is 30", in part because Coleman promoted this width and also because 30" can be easily straddled by most people. We use 42"-wide beds because it gives us an additional 12" of growing space. In our greenhouses, our bed widths are wider—54"—to maximize space even more. Optimum bed width is up to you, but in general wider is more efficient as long as you can easily step across the growing surface.

### 3. Choose a bed length between 50' and 100'.

Ideal bed length is also subjective, but a good rule is to stay between 50' and 100'. Beds longer than 100' require too much walking from end to end. Beds shorter than 50' aren't big enough for efficient production. Important note: Keep all beds the same length so that row covers, drip tapes, and tarps can be used interchangeably.

If you are designing a farm and think you may eventually use a paperpot transplanter, I recommend 75'-long beds. Beds at our new farm are 75' in order to minimize waste of the paperpot chains: A 42"-wide by 75'-long bed accommodates two chains at 6" spacing. We grow many of our core crops (such as cilantro, basil, and lettuce) using this system, as I describe in appendix A. Also,

Overhead view of a Clay Bottom Farm plot. Tiny-farm plots should be small enough to easily access crops but large enough for sufficient production. Photo courtesy of Adam Derstine.

our largest greenhouse is about 150′ long, so when that growing space is divided in half, bed length in the greenhouse matches the field.

### 4. 12″ pathways are best.

In the deep-mulch system, compost is spread uniformly across the entire plot. Thus, paths between growing beds should be as narrow as practical. Most anyone can walk on a path 12″ wide. The one exception to this rule on our farm: For greenhouse tomatoes, we widen paths to 24″ to accommodate a harvest cart.

### 5. Slope beds and pathways downhill.

When laying out beds, take slope into account. You want to use the 12″ pathways between plots as drainageways—water should run off the beds, into the paths, and away from the garden. For optimum drainage, slope beds and pathways downhill. If beds are oriented the wrong direction, across the slope, water can collect in the paths during heavy rains.

## A Simpler Way to Manage Pathways

I recommend leaving pathways clear, without adding mulches like straw, grass clippings, wood chips, or cardboard. At the scale of a home garden, wood chips and other mulches create a tidy appearance and, for many, work well. However, on a market scale, these mulches make it more cumbersome to remove weeds when the mulches break down. Also, mulching with these materials adds work and cost. Remember, simple is best.

We've found that the most efficient approach to pathway management is to shallowly cultivate them early and often—before you even see weeds—with Hoss fixed-blade sweeps attached to a wheel hoe. The Tilmor E-Ox electric wheel hoe also works well for this job if your pathways are wider than 16″. I love this tool already, and I have used it for only a season. I describe these tools in chapter 8.

To initially establish pathways, we use an aluminum scoop shovel and walk backward, pulling the shovel along to create a smooth path. The goal is to mark an area for workers to walk, not to create a raised bed.

### 6. Use walkways as drainageways.

Likewise, walkways—the mowed areas around your plot—should be designed with drainage in mind. Walkways should be slightly lower than the garden plot. This will happen naturally once you add 4″ of compost to the plot.

If your walkways lack adequate slope or slope the wrong direction, I recommend tilling them up and reshaping them. In 2021, we tilled a walkway in the fall, reshaped it so that it carried water away from the garden, and seeded it with grass in early spring. We use 6′-wide grass walkways between our plots, sufficient for driving a golf cart.

### 7. Permanently stake the corners.

Essential tip: Pound permanent steel or wood posts into the four corners of your plots once you have them staked out. Plot boundaries tend to shift with time, and permanent corners will let you reestablish the edges as needed.

### 8. Edge every season.

There are three reasons to keep a clean edge between garden and yard. First, a clear demarcation helps you know where to spread compost. Second, a clean edge keeps weeds from creeping in. Third, it just looks better. Every season, we string mason line between our four corner posts and edge our plots with a battery-powered edger (see page 156). This can also be done by hand with a straight-bladed shovel but will take much longer.

# Getting Started with Deep Mulching at Clay Bottom Farm

A few days after we bought our new farm, I pounded in wooden temporary stakes to mark the corners of our two growing plots and of our first greenhouse. Then I tilled the plots about 4″ deep with a 34-horsepower Kubota tractor and 60″ tiller attachment in order to break through the pasture grasses and other perennials. In retrospect, plowing first would have done a better job because the pasture was full of deeply rooted, stubborn perennials. A few butterfly weeds and Canadian thistle continue to push their way through our deep mulch. We pull these out by hand as soon as we see them in the spring.

Then we added compost from our old farm to these new plots, topped with compost that we purchased from a local environmental center. Working piecemeal but methodically, it took two seasons to eventually cover all growing surfaces with at least 4″ of compost. In a few places, to even out the grade or build up extra-sandy spots, we covered the soil 6″ to 8″ deep with compost.

The first plot that we covered was the ground on which we planned to build our first greenhouse. By laying compost on the surface, we were able to build up the ground inside the greenhouse so that water drained away easily.

Next, we built a fence whose posts mark the boundary of our growing area. The fence is made with salvaged wooden posts, spaced about 20′ apart, with a line of electrical horse tape spaced 6″, 12″, and 18″ off the ground to discourage groundhogs and 4′ and 7′ off the ground to dissuade deer. The fence is not fool-proof. An occasional deer or groundhog will find its way through, but the fence (and our farm dog, a cattle dog) stops most of them. The electric charger is plugged into a timer that turns off the fence during the daytime. We mounted a switch that controls power to the fence near the door to our house, so we can easily turn it off manually as needed. Proper fencing is an excellent way to prevent waste.

The posts of this fence also mark the corners of the plot. This electric fence is only energized at night. We can easily step through it to access plots.

The biggest mistake we made, as we built our plots, was to underestimate the amount of drainage they would need. During the first year on our new farm, Goshen experienced a 100-year flood. Many homes and businesses were ruined, and water ran through our farm for days. The deep compost soaked up rain like a sponge; but in a corner of one plot, the water collected because it had nowhere to go. Plant roots became waterlogged, and many died. In response, we reshaped pathways to carry water away faster. We also more intentionally shaped the pathways and connected them together, forming a sort of shallow canal, to direct water around greenhouses and plots to a retention pond that we dug at the south end of our farm. In that low corner, we added another 2″ to 3″ of compost. Now, we don't have water problems in that corner.

In summary, to start high-production beds quickly, lay down thick compost made from local ingredients, and let biology get to work. This will simplify your fieldwork for years to come. With this one step, you suppress weeds, hypercharge the soil biosphere, and supply plants with nutrients for many seasons. Deep mulching is the gateway to a high-profit micro farm.

# CHAPTER 7

# Two-Step Bed Flipping: A Method for Increasing Your Production on a Small Footprint

As I discussed in part 1 of this book, to essentialize means to discern which activities are truly necessary, or vital, and which ones aren't. When preparing garden beds, there are only two things you really need to do—clear away old crops (or weeds) and smooth the ground. Everything else is probably superfluous, or trivial, to use Pareto's term. In this chapter, I explain a simple way to essentialize bed preparation.

On tiny farms, market gardeners might flip—or transition—beds from one crop to another several times per season to make the most of every square foot. With these rotations, you effectively double or triple your production area by planting the same beds two or three times each year. That is, a ⅓-acre plot literally becomes equal to a ⅔- or 1-acre farm growing one crop per year. This justifies applying a large amount of compost to keep fertility levels high for multicropped beds.

The key is to flip the beds quickly and efficiently. At Clay Bottom Farm, we flip more than 100 beds each season—easily our most burdensome task, and an opportune place to lean up.

Here is the bed-flipping process we devised when we moved to our new farm. First, we cover old crops (usually without cutting) with a silage tarp and let the greens decay in the sun (step 1); within a few weeks, the crops will turn to mush. Next, we remove the tarp and rake any debris into the path, if needed (step 2). The bed is then ready to replant—no broadforking, chisel plowing, rotovating, hoeing, undercutting, or fertilizing required. The method is as uncomplicated as it gets.

The rule guiding the process is to leave roots in the ground. Roots left in place provide carbonaceous material for soil microbes to eat. Also, as they decay, roots leave air pockets behind, loosening the soil structure. The decayed above-ground plant matter also feeds the soil.

This method extracts the farmer from the growing process—you do just enough, as it were, and then walk away while nature does the rest. We save hours every season with this approach, compared to conventional methods like tilling, using a power harrow, or pulling out old crops by hand, while sparing our bodies unnecessary hard work. After 16 years of growing, I've concluded that most methods promoted for bed preparation, like those just mentioned, can be skipped, especially if you are deep mulching. They don't pass the Pareto test.

As with deep mulching, this method would be impractical in a large operation with acres in production—it would be too cumbersome to manage the tarps. Below I describe two-step bed flipping in detail.

## Just Smother It: A Better Way to Flip Short Crops

For spinach, baby greens, head lettuce, and other "short" crops, we pull a 14′ × 75′ silage tarp over top of the beds when we are done harvesting them. This size of tarp covers two beds at a time plus the paths on either side. If we need to cover just one bed, we'll fold the tarp in half. I recommend cutting tarps to narrow widths like this to make it easier for one person to handle them. Remember Schumacher's advice: Design with humans in mind. Make your farming comfortable to do.

Cut tarps a bit wider than the bed + path width you plan to cover, as the crops underneath will push the tarp up slightly. Ideally, tarps are wide enough to cover the paths on both sides. Silage tarps for market gardening are available many places online. We buy ours at a local farm supply in order to save shipping costs, which can be substantial due to the weight of the tarps.

Next, we pin down the tarps with sandbags, which we fill with driveway gravel for optimum drainage. One sandbag spaced every two paces—about 6′—keeps the tarps pinned down in the wind. If you are in an extra-windy area, choose a closer spacing. If we are using more than one tarp side by side, we make sure to overlap the edges by at least 12″.

*Tip #1:* Align the edge of your tarp precisely with the edge of the plot. If the tarp is misplaced even by a few inches, you've given weeds room to grow. They will take full advantage of the opportunity.

Covering a bed of spinach with a silage tarp. The spinach is not cut before-hand. In two weeks, we will be able to pull back the tarp and rake the bed, leaving the roots in the ground. Note that the tarp is big enough to cover two beds at a time.

*Tip #2:* Stretch the tarps a bit as you lay them so that they are as tight to the ground as possible. The tarps kill the crops through light deprivation—occultation—and by trapping heat. The tighter your tarps, the darker and hotter it is underneath them.

*Tip #3:* If you've laid down drip tapes to irrigate, leave them in place under the tarps. They are easier to remove, if needed, after the crop has died.

*Tip #4:* Farm like a tree and work ahead. Tarp your beds at least two weeks ahead of when you need to plant. Don't wait until the last minute. On our farm, at least two beds are usually covered by tarps at all times in preparation for planting.

There's no reason to mow, cut, or shred crops before covering them, unless you want to trim them down so that the tarp lays flatter. For example, we will sometimes trim cilantro and basil plants if they are knee-high or taller. We leave these trimmings on the bed surface, under the tarp, to decompose in place. In most cases, however, it's better to save your time. Worms and microbes will sufficiently shred and chew up uncut greens.

After the greens have "melted" (died) under the tarp, we pull it off and rake the decomposed matter off the bed and into the pathway, clearing the bed just enough to plant or transplant into it. If transplanting by hand, we

This bed of green beans was covered with a tarp and then raked a few weeks later, clearing room for the next planting while leaving roots in the ground. The bed will not be tilled before planting, though we might loosen it with sweeps or forks, if needed.

On occasion, we will trim down gangly crops before covering them, to help the tarp lay flatter. Here we are cutting cilantro with a weed eater before covering with a tarp. A scythe also works well for this job. Note that adjacent beds are temporarily covered to protect them from bits of cilantro.

will sometimes leave the residues on the surface as a kind of mulch instead of raking them off. However, raking is usually preferred to make transplanting easier.

Occasionally, this mess of decomposed matter is too bulky to fit neatly in a path. In this case, we rake it into a pile at the end of a bed, creating a small compost heap that we will spread across the bed in a month or two, after the greens have more fully broken down. Alternatively, we might rake these spent greens underneath another nearby tarp—out of sight, out of mind—or add them to a compost pile near the plot. The point is to keep old plants as close to where they grew as possible, to let them enrich the soil and save yourself a backache.

There are rare occasions when we tarp a crop even before it has

HOME GARDENER MODIFICATION
## How to Flip Beds in a Home Garden

Two-step bed flipping can easily be done in a home garden. Ideally, try to locate a large piece of black plastic, perhaps a scrap from a local farmer. However, anything that will block out sun can be used to smother crops—big sheets of cardboard or tarps work just as well. Or perhaps you have access to leftover construction materials, like rubber roofing scraps. You can also buy landscape fabric, which works for bed flipping, from most hardware stores. Or smother old crops with black plastic garbage bags that you cut open at the sides to create a long rectangle. You can pin down your smothering sheets with any kind of weights that are handy, like cinder blocks, buckets, or bricks.

finished maturing. For example, in 2021, a bed of lettuce was overtaken by fast-growing Peruvian daisies that we had mistakenly let go to seed in 2020. We decided that instead of fighting the weeds, we would cover the bed with a tarp, let the daisies and lettuce melt, and start over in a few weeks with another crop.

## HOW WE STORE SILAGE TARPS

When a tarp is ready to be removed from a bed, we pull it onto another bed of harvested greens. If no beds are ready for a tarp, we store it in the field, flat and open, on top of another tarp. This avoids the need to fold the tarps and move them out of the field.

## DOESN'T TARPING KILL MICROBES?

Short answer: Yes, many (but not all) microbes will die under a tarp, especially in midsummer, but they repopulate quickly.

According to researchers at Texas A&M University, the optimum temperature range for most soil microbes to thrive is between 60 and 85°F (16–29°C).[1] Many fungi, bacteria, and other microbes will die when temperatures reach 145°F (63°C).[2] However, some thermophilic ("heat-loving") microbes thrive in soils that reach 150°F (66°C). In the heat of summer, adding a tarp can warm soils above 130°F (54°C), even in the north, thus tarps do slow down microbial life. However, bacteria and protozoa can repopulate within hours and nematodes within 30 days.[3] The effect of

## Clear Plastic vs. Silage Tarps

Using clear plastic over soil is called *solarization*; using opaque plastic of whatever color is called *occultation*. While most growers use black plastic, there might be occasions to use clear plastic. Research shows that green matter dies faster and soils heat more quickly with clear plastic compared to colored plastic *if* weather conditions are hot and sunny. Thus, clear plastic (such as used greenhouse film) can be the better choice to kill weed seeds in midsummer. During colder and darker seasons, however, clear plastic can actually encourage plant and weed growth because it allows the sun to pass through it but lacks the opacity to kill emerging plants. As such, black plastic is the safest and most versatile year-round choice.

In our experience, black woven landscape fabric can also work for bed flipping, but because it is porous, plant debris can take a bit longer to break down than it does under silage tarps.

tarps on soil microbes is an area of emerging research, but so far scientists agree that to speed microbial rebound, you should keep the soil full of carbon (another reason to leave plant matter and roots behind) and moist but not saturated.

We kept the soil moist under this tarp using drip tapes in order speed the recovery of biological life.

### WHY DO WE IRRIGATE UNDER TARPS?

If beds are dry, we moisten them with drip tape irrigation or with overhead sprinklers before placing tarps over them. If we plan to tarp for a long time, we will place at least two runs of drip tape across each bed, under the tarp, and irrigate the bed for the entire time a tarp is in place. We use timers on our hydrants to regulate soil moisture. We test how much to irrigate using a finger—at 2″ deep, the soil should feel moist but not sticky.

We irrigate under tarps for two reasons: First, as stated in the previous section, microbes will rebound faster if the soil is moist. Second, compost is hydrophobic—that is, it is difficult to resaturate after it dries out. Keeping the deep compost moist prevents this problem.

## HOW LONG DOES IT TAKE FOR A TARP TO TERMINATE A CROP?

Decomposition time depends on the size of the plants and the weather. In mid-July, small leafy greens like arugula, covered with a tarp, will turn to mush in a few days. On the other hand, full-size Tokyo Bekana cabbage, covered in the fall, might not fully die until spring. We've found that from May through September—the season during which we flip beds most often—two weeks is usually sufficient for most crops to decay.

## WHAT IF I DON'T HAVE TIME TO WAIT?

If you have seedlings ready for immediate transplanting or if you are in a hurry to direct seed, the best approach is to remove the old crop by hand. First, we loosen roots by slicing beneath the soil surface with fixed-blade sweeps, which we buy from Hoss, attached to a wheel hoe. This is a good

If we are in a hurry to replant, we will slice underneath an old crop with a wheel hoe and Hoss sweeps before raking the bed by hand. This is a good job for two people. The Tilmor E-Ox electric wheel hoe can also perform the job with one person.

job for two people—one person pulling and the other guiding the tool. To help the tool dig in, we might add a sandbag or two to the middle of the wheel hoe. Alternatively, the new E-Ox electric wheel hoe from Tilmor can also be used, with the Hoss sweeps, allowing one person to complete the task. We then rake the greens into a pile, using adjustable spring tine rakes, and we carry them off with pitch forks to the compost pile.

## HOW TO PREPARE BEDS FOR WINTER

Just as trees cover the ground under their canopies with leaves each fall, it is a good idea to cover your growing beds with an inch of compost (or slightly broken-down leaves) before winter. This mulch will act like a sponge, reducing the problem of soil erosion.

If beds have greens on them going into winter, we leave the greens in the field—their roots help to "hold" the ground to prevent erosion, and freezing temperatures will kill the tops, which will be ready to rake away in the spring. In our experience, deep-mulch growing beds don't need to be covered with tarps or with cover crops in the winter, though these practices don't hurt. On our farm, as stated, there is not enough time to establish cover crops, because the farm is almost always filled with cash crops.

## Flipping Tall Crops

Some crops, like full-size kale and trellised cucumbers and tomatoes, obviously can't be easily covered with a tarp. Instead, we cut these crops with a pruning lopper or a hand pruner as low to the ground as possible so that stubs don't interfere with future plantings. No tarping is necessary following this cutting, as long as the plant is cut low to the ground. As I'll explain further in chapter 10, we grow kale, cucumbers, and tomatoes into landscape fabric with holes burned into it. We remove the landscape fabric *after* cutting the spent crops. Then we rake if needed and replant the bed.

Cutting full-size kale with a pruning lopper. Later, we will remove the landscape fabric and plant around the old roots that were left in the ground.

## DO ROOTS IN THE GROUND
## DISRUPT SEEDING/TRANSPLANTING?

When we seed or transplant on a no-till bed, we stagger rows to avoid hitting the stubs and roots from the previous crop. For example, let's say a bed contained three rows of basil. Following such a bed, we would seed five rows of spinach, with the old basil roots in between rows. Old roots

Here we are transplanting lettuce into a bed where basil grew previously. Note the stubble from the basil. In our experience, old roots left in the ground do not negatively affect growth of a new crop.

left behind, in our experience, have never negatively impacted the growth of new crops. Also, in our experience, the Jang seeder and the paperpot transplanter push their way through old roots as long we stagger the rows as described.

## IS IT OK TO STEP ON BEDS?

A bit of treading on beds is just fine. There are many times when stepping on a bed saves effort, such as when harvesting greens or when scurrying across a plot sideways. Soils with plenty of compost can support more weight than other types of soil.

That said, it's best to stick to paths when practical in order to preserve the soil's structure and porosity. And it's never a good practice to step on beds when they are very wet because wet-compaction can take a long time and considerable effort to repair.

## DO I NEED TO ROTATE CROPS?

We do not intentionally rotate crops on our farm. We've grown long-season kale in the same location for three seasons in a row, and we frequently follow lettuce with lettuce, tomatoes with tomatoes, and so on. As long as we apply a gentle approach to soil management—not tilling and applying compost to the surface every few seasons—we have not seen adverse effects from this practice. In fact, we wonder if microorganisms might develop multiseason symbiotic relationships with plant roots from the same species. In any case, on a tiny farm, we save a lot of planning time and simplify our work by not concerning ourselves with crop rotation.

## DO I NEED TO LOOSEN
## THE SOIL BETWEEN CROPS?

We have found that there is rarely a need to loosen soils between crops. The compost, the fungi, and the roots left behind by old crops keep soil sufficiently loose for most crops. There are a few occasions, however, when we will loosen soils with tools—for example, to prepare beds for fall carrots, which can grow 12″ in length. These carrots penetrate below the layer of deep mulch, and we have seen longer carrots in loosened soil. In spring, if the ground was compacted by snow, we might use our electric wheel hoe to slice 4″ under the surface, lifting it slightly. Mostly, however, we take a less-is-more approach to soil loosening and simply plant right after raking.

## Testing for Biological Activity

A major benefit of two-step bed flipping is that the soil biota stays active. Thankfully, many labs are beginning to develop methods for testing biological activity in soils, not just nutrient levels. Due to the complexity of soil life, however, this is not a straightforward task. Agronomists typically weigh a combination of factors to determine "biological health." These can include:

1. The number or percent weight of roots in a given amount of soil.
2. Available organic matter.
3. Soil aggregation (how well the soil holds together).

4. The amount of $CO_2$ given off by a soil, as measured by a Solvita respiration test. $CO_2$ is a by-product of decomposition; thus, it indicates how actively microbes are working.

On our farm, we use a local testing facility, A&L Great Lakes Laboratories, once per season to give a quick, low-cost measure of nutrient levels and organic matter. For a deeper read, we also have used a Haney soil test, available online, which measures $CO_2$ and soil aggregation. On both types of tests, numbers consistently improve when we keep compost in place and employ a gentle approach to soil management.

Beans and lettuce planted in no-till beds.

Our driving motivation in essentializing fieldwork is to cut out the superfluous steps to free up our time. When preparing beds, you really only need to terminate the old crop (by tarping or cutting it at the base) and then rake. Through deep mulching and two-step bed flipping—two streamlined micro-farm methods—we've shaved hours per week compared to before, work less hard, and enjoy life more on and off the farm. In the next chapter, I'll describe a few good tools that help us get it all done.

# CHAPTER 8

# Seven Vital Tools (and a Few Others)

In early 2018, as we built our new plots and began construction on our greenhouse and barn-house, we were also starting to pack up the old farm, deciding what to bring with us and what to leave behind, give away, or sell.

This should have been an easy task, but it wasn't. Despite years of leaning up, our old farm was still full of too many things. This move was another opportunity to cut back.

We sold our old farm in the spring, but we retained possession of our growing plots until the fall so we could harvest crops that we'd already planted. So that summer, we farmed in two places. I drove back and forth between farms each week, farming mostly with tools that I threw into the back of our van. It dawned on me, after a few months, that those were the only tools we really needed.

Earlier, I made two assertions about tools. First, that it is best to farm with the fewest possible number of tools, and second, that tools should have a "human face." These rules are especially true on a micro farm. Farm lightly with a few appropriate tools, and you set the stage for productive growing.

## Starter Kit Tools

These are basic tools that I recommend for micro farming. They add a lot of value at a relatively low cost. They are all you really need to get started. In fact, I also recommend this starter kit for home gardeners, except for the wheel hoe. If you have a very tiny plot, a wheel hoe might be unwieldy.

## SEVEN VITAL FIELD TOOLS

As stated earlier, we accomplish most of our work with just seven field tools. In order of how frequently we use them, here are the field tools I kept throwing into the back of the van along with a word about how to use them.

### 1. 30″ bed preparation rake

This rake is distinguishable from a standard garden rake because the angle of its tines can be adjusted to effectively "skim" beds of debris while standing in a comfortable upright position and because it is lighter in weight. Its tines can glide collinearly with the soil surface, making it an ideal tool for no-till bed preparation. A standard rake would dig too deeply. It also features a longer handle than most rakes.

We use this tool almost daily to clear beds of debris and to level them before planting. The rake is available at Johnny's Selected Seeds, and a heavier-duty version is sold through Earth Tools. Johnny's also sells row markers, tubes that can be placed over selected tines to aid in marking rows. Or you can cut your own tubes from scrap ½″ PEX plumbing pipes.

### 2. Fixed-blade sweeps on a wheel hoe

The fixed-blade sweeps from Hoss Tools can be adjusted to fit on many types of wheel hoes. Our wheel hoe is from Glaser. The blades are sturdy and "fixed," meaning they don't swivel like an oscillating hoe. Power from your upper body is transferred directly to the soil.

We use the sweeps to cultivate pathways, and we rely on them to loosen compacted ground, a job typically reserved for broadforks. We've found the sweeps to be more ergonomic and faster to use than broadforks. We also use

Bed preparation rake with tubes for row marking.

Adding fixed-blade sweeps is a way to upgrade a wheel hoe.

this tool to slice underneath roots to loosen them.

### 3. Aluminum scoop shovel

The lowly scoop shovel earned its coveted spot on our tool list because it's multifunctional. We spread compost with it and use it to grade pathways. Scoop shovels are available in a variety of widths, and I recommend buying one whose front edge matches the width of your paths between growing beds.

We winnowed our field tools down to these seven. They hang in the greenhouse in the winter and on a rack near the garden in the summer.

### 4. Adjustable-width rake

This low-cost rake features tines that can be adjusted between 6″ and 24″ wide. We bought ours at a local hardware store. We use it to clean pathways, rake debris from growing beds, and for general yard cleanup.

### 5. Half moon hoe

This classic European tool is both powerful and precise. Its sturdy gooseneck is strong enough to allow for deep digging, and its fine points scuttle with finesse underneath and between plants. In other words, it is one hoe that performs the work of many, perfect for a minimalist. We have tried dozens of hoes, and this is still the best all-around hoe, in our experience. Its head is 6½″ wide. Our half moon hoe is made by DeWit.

The adjustable-width rake is one of our most-used tools.

### 6. Narrow collinear hoe

The Johnny's Selected Seeds narrow collinear hoe, designed by Eliot Coleman, outperforms other small hoes that we have used. With a 3¾″-wide

If you own just one hoe, I recommend the half moon hoe because it is ergonomic to use, durable, and can reach weeds in tight spaces.

145

The narrow collinear hoe fits perfectly between rows of closely spaced greens.

head, it is just the tool to cultivate between greens seeded closely together. It glides on the surface, collinearly, instead of digging into the ground. This makes it much easier to use than standard hoes.

### 7. Garden fork

Our Clarington Forge garden fork is the most durable digging fork we have used. It is perfect for loosening carrots and potatoes, dislodging stubborn weeds, or moving straw. We broke many cheaper digging forks before splurging on this one, which should last a lifetime. The fork is made from one piece of steel that extends far up the handle, past the point where our previous forks always broke.

## TWO HARVESTING TOOLS

When we moved, we trimmed an array of harvesting tools down to two.

### 1. Curved grape shears

The curved tip on this tool gives it an edge, so to speak—it allows you to harvest without twisting your wrist awkwardly around fruit clusters to reach to

Knives and pruners for hand harvesting are stored on magnetic bars near the entrances to our processing room and greenhouses.

the stems. This seems like a minor point, but after harvesting 500 pounds of tomatoes, your wrist will appreciate that little curve. That said, many types of pruners can work to harvest tomatoes, peppers, and the like, and I think the best pair of shears is one that fits comfortably in your hand.

### 2. A 6″ stainless steel restaurant-grade produce knife

We use this knife, designed for use in restaurants, to harvest baby greens. We own several, and we hang them on magnets attached to our greenhouse and outside the door to our processing room, where they are quick to grab. The square edge on the knife makes it safe to handle. Also, the tool sanitizes easily because of its simple shape. It is widely available through online restaurant retailers.

# A Simpler Germination Chamber

A germination chamber is an insulated box used to start seeds. It is the most reliable, low-cost way to germinate seeds, obviating the need for heating mats, heated benches, and the like. It is an example of a simple and productive solution to a complex problem.

We constructed ours from an upright freezer gifted to us by a neighbor. The freezer has a large door for easy access. An old slow cooker, plugged into an INKBIRD external thermostat, provides heat and humidity. The slow cooker sits on the floor of the freezer and is filled with water. We taped the probe from the thermostat along the back wall of the freezer box.

In summer, the freezer's compressor—also controlled by the INKBIRD—can cool the chamber for germinating lettuce and storing ripe tomatoes.

After a few days in the chamber, seeds will "pop," or show a cotyledon just above the soil surface, indicating that it's time to move trays to the

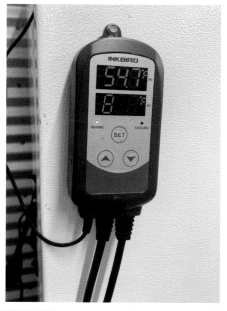

We start almost all seeds in this germination chamber, made from an upright freezer, and controlled with an INKBIRD external thermostat.

## What Is the Best Way for Home Gardeners to Germinate Seeds?

Even on the scale of a home garden, I recommend using a germination chamber to start seeds. The simplest home-scale germination chamber is a large picnic cooler. You can put a small slow cooker in the cooler, hooked up to an INKBIRD thermostat, or simply turn the slow cooker on for a few minutes each day and keep tabs on the temperature. This setup is low cost (potentially free) and more effective than electric heating mats. Remember to check on your plants every day, and remove them to a south window or (better yet) under grow lights as soon as the seeds pop. For a simple grow light setup, see "LED Grow Lights," on page 168.

grow-light table or to the propagation house. We have had nearly 100 percent germination success with this system since we first put it to use.

To save room inside the chamber, we stack trays right on top of each other. This practice has never presented a problem, as long as we check the chamber twice per day and pull out trays before the seedlings become leggy.

# Next-Level Tools

I recommend the tools in this section if you plan to grow professionally. We use each of them frequently. They are purposely designed for micro-farm use. We chose them carefully because they shave time and are durable in their construction; they pay for themselves. I call them "next-level," however, because they are more costly and because lower-cost alternatives do exist. Many of these tools might be a good investment after a season or two.

## ELECTRIC-POWERED UPGRADES

Lean, small farming isn't about turning back the clock or adopting a Luddite mindset. Sometimes, it makes more sense to plug in.

### 1. Tilmor E-Ox wheel hoe

To ease burden, we sometimes use a battery-powered wheel hoe, the Tilmor E-Ox. This tool is a good example of technology with a human face. It is both innovative and purposely designed for use at a human scale.

Using the Tilmor E-Ox electric wheel hoe with Hoss sweeps to cultivate a path.

We use the tool to perform the same work as our wheel hoe, with considerably less effort.

The E-Ox is supremely versatile. It is powered by two 20-volt power-tool batteries and can accept batteries from a variety of manufacturers, including DeWalt, Makita, and Milwaukee. It can accommodate attachments designed for Planet Jr., Glaser, and other wheel hoe manufacturers. Front and rear tool hitches also let you attach rakes and other tools. It offers five speed settings, controlled by a handlebar.

I have a personal preference for nonpowered tools, but as I age, I find that I appreciate a battery boost now and then. Tools like the E-Ox also make farming more accessible for people with disabilities. As described in chapter 7, we outfitted our E-Ox with Hoss sweeps, and we primarily use the tool to help remove old crops (it does a beautiful job of slicing underneath roots). We also clean walkways with it, though because the E-Ox has a tire spread of 16″, it doesn't easily fit on pathways wider than that distance.

The E-Ox is similar to the Tilther, a battery-powered small tiller sold through Johnny's, but in our experience the E-Ox is better. It works in a wider range of soils without getting plugged up by rocks, as sometimes happens with the Tilther, and it is gentler on the soil.

## 2. Quick-cut greens harvester

We use this tool—another example of appropriate, human-scale technology—several times each week to harvest baby greens; it replaces the produce

## Tools Should Be Comfortable to Use

A few weeks after I turned 40, I came into the house one day with a stiff back. I expected the pain to subside in a few hours, but it hung on. I visited the doctor, who basically told me, "Welcome to the over-40 club." She sent me home with a page of stretches to do and a bottle of enhanced Tylenol, and she suggested I figure out more ergonomic ways to work. I had no interest in relying on pharma-ceuticals to ease my pain, but her advice did inspire me to extend the handle on our greens harvester so I could use it from an upright position. The original tool required bending over in an awkward way. A good tool shouldn't be awkward to use. It should feel good in your hands—and not break your back. This is what it means for a tool to have a human face.

knife. There are many greens harvesters on the market, but because of its portability, the Quick-Cut Greens Harvester, made by Farmers Friend, is particularly suited to tiny farms. It can be powered using any cordless drill, but it's best to use a three-speed drill with this tool so it can operate at a variety of speeds.

We modified our greens harvester by replacing the manufacturer's handles with 36″-long aluminum wiggle-wire extrusion left over from a greenhouse project, which allows me to harvest from an upright position. This modification takes just a few minutes and requires tools most farmers have on hand. The idea for this modification came to me as I laid in bed one afternoon from a sore back. As of this writing, the tool is being redesigned by Farmers Friend so that a taller version is available more widely.

### THREE CARTS TO LIGHTEN THE LOAD

A market gardener might get by for a season or two carrying produce by hand, using baskets like the Gorilla Tubs sold in many catalogs. But eventually, you'll want professional-duty carts to take the load off. In midseason, we haul thousands of pounds of produce from greenhouses and the fields to the processing room every week. Here are the carts that we use.

#### 1. Vermont garden cart

This classic cart is a favorite on almost all tiny farms I've visited, because it is purposely designed for market gardeners and serious home gardeners—with people in mind, as Schumacher might say.

I extended the handles on this greens harvester using leftover pieces of greenhouse wiggle wire channel to make it more ergonomic to use. Photo courtesy of Caleb Mast Photography.

## When You Can't Buy It, Make It Yourself

On occasion, it is better to build your own tools to save money and to design something just right for your farm. We built this simple portable root washer out of leftover bicycle parts and a greenhouse benchtop. Larger industrial models didn't fit our scale—we couldn't move them about the farm—or our budget. A good book with ideas for home-built tools is *Build Your Own Farm Tools* by Josh Volk.

A group of visiting students wash carrots in our carrot washer.

The primary feature we appreciate: Its tires are spread far enough apart to straddle growing beds. It also holds an incredible 500 pounds of weight; it distributes the load over the tires, which are large in diameter, instead of the handlebars, so one person can push and lift an entire load. The large tires also make it easy to roll the cart over rough terrain. Its high sides are ideal for piling in leaves, and its removable back is just the thing for dumping them.

An electric golf cart is indispensable on a tiny farm.

### 2. Electric golf cart

For most hauling on our farm, we use an electric golf cart. These are readily available on the used market as golf courses replace their fleets periodically. The golf cart is ideal for moving around sandbags, field tools, T-posts, and ladders. We outfitted ours with an aluminum utility box that opens at the back. Golf carts are compact and fit the scale of most tiny farms.

Rachel deserves credit for pushing us to buy this vehicle. While I often want to tough it out, she has a better instinct for making jobs easier to do. Now that we own a golf cart, it's part of the family. I can't envision parting ways.

### 3. Flatbed tomato cart

This cart has one purpose: to aid in harvesting tomatoes. With a 24″ profile, it fits perfectly in the pathways between tomato rows. We put boxes on the cart and fill them with tomatoes as we pull the cart along behind us. In fact, as we shape beds for tomatoes, we do a test run up and down paths with the cart to make sure it can round the corners. If you plan to grow tomatoes, I recommend a flatbed cart.

## FIELD TOOLS FROM KOREA AND JAPAN

Our most sophisticated but still simple-to-use field tools are a seeder from Korea and a transplanter from Japan. These countries, where small farmers play a more meaningful role in food production than in the United States, have developed better tools purposely designed for micro farms.

When we started using these tools in the mid-2010s, few tools from Asia were readily available in North America. Now companies like Niwaki and Japan Agri Trading, among many others, have made tools like these widely available.

The Jang seeder easily glides through rough soil.

### 1. Jang JP-1 Seeder

This is the best seeder for use in a no-till system because it glides easily through rough soil. It does not require perfect conditions. Teeth on the front wheel can dig into the ground, and a rear press wheel packs soil around seeds. While the tool is available in multi-gang configurations, it's best on

**Table 8.1.** Clay Bottom Farm Roller Sizes for Jang JP-1 Seeder

| Crop | Roller Model | Brush Setting | Inches Between Rows |
|---|---|---|---|
| Carrots (use raw seed) | F-12 | Low (keep the brush just off the roller) | 8 |
| Baby lettuce | F-12 | Low | 5 |
| Baby Asian greens | X-24* | Low | 5 |
| Spinach | LJ-24 | Low, medium, or high (just under the height of the seed), depending on seed size | 6 |

Note: Sprocket setting is always 13-front, 11-rear
* For thick stands, seed twice in the same row.

**Table 8.2.** Clay Bottom Farm Paper Pot Chart

| Crop | Gravity Seeder Top-Plate Size | Seeds per Cell | Notes |
|------|-------------------------------|----------------|-------|
| Basil | Hand-seeded | 4–6 | |
| Cilantro | 4 mm | 4–6 | |
| Head lettuce | 4 mm | 1 | pelleted seed only |
| Spinach | 4 mm | 4–6 | double seed if required |
| Turnip | Hand-seeded | 3–4 | |
| Spring onions | 4 mm | 1–2 | |

Note: Paper chain size is 6″ except for spring onions, which use 2″ chains. Crop spacing is always 9″ between rows.

a tiny farm to stick with the single-hopper version.

Table 8.1 shows the roller and sprocket sizes that we use for various crops in the Jang seeder.

## 2. Paperpot transplanter

If you transplant for several hours per week in the springtime, I recommend a paperpot transplanter. We have used the paperpot system for more than eight seasons. I wouldn't think of farming without it, even at our tiny scale. We grow more than half of our crops with the tool. Table 8.2 is the chart we developed to simplify its use. In chapter 10 and appendix A, I show specifically how we use the tool on our farm.

Planting Genovese basil with a paperpot transplanter.

## NEW LITHIUM-ION TOOLS

In the past decade, lithium-ion battery technology has developed rapidly as engineers find ways to put more power into smaller and lighter battery packs. The E-Ox and the greens harvester are examples of lithium-ion-powered tools. As time goes on, more electric-powered tools, purposely designed for serious home gardeners and market gardeners, will surely become available. We are always on the lookout for them. Here are three more lithium-ion tools that we frequently use.

### 1. Universal-attachment brush cutter / string trimmer / edger

This tool, from DeWalt, supports many types of attachments. The brush cutter cuts off tall, thick-stemmed plants like basil at the base. The string trimmer is useful for mowing greens low to the ground and for cleaning up tall grasses around our plots.

The edger attachment creates a sharp line between mowed areas and our plots, which prevents grasses from forcing their way in. Edging plots is critical on a no-till farm. Previously we would run our tiller around the perimeter of a plot as a kind of edger. Without the tiller, we use the edger to reestablish borders. Clean lines of demarcation between plots and pathways help tidy the farm, and they indicate where to place tarps, set up drip tape systems, and seed crops.

This lithium-ion edger, from DeWalt, features interchangeable heads. We also use the string trimmer and brush cutter attachments.

### 2. Blower

We keep a blower handy to clean the floor of our hosing area, to blow out cobwebs on our porches, and to clean the floors weekly in our summer greenhouses. (Greenhouse floors in summer are usually covered with landscape fabric.) We also sometimes use it to blow "bits"—pieces of cut greens—from baby greens beds after harvest. Our blower is also made by DeWalt, so its battery packs are interchangeable with our other DeWalt power tools.

### 3. Jacto PJB backpack sprayer

This lightweight sprayer, another quality tool from Japan, uses a lithium-ion battery to pressurize solutions for spraying. We use it to sanitize landscape fabric that we use from one season to the next with OxiDate, and to spray DiPel DF, an organic insecticide, on tomatoes when hornworms appear and on kale to prevent cabbage loopers. The Jacto sprayer is more precise and powerful than similar-sized models that rely on a hand pump for pressurization. A potentiometer allows the user to adjust the spray strength, and the nozzles allow for a variety of spray patterns. For a photo of the backpack sprayer, see page 206.

## Caring for Tools

To keep tools in top condition and ready for use, we've put in place routines for their care. Annually, in winter, we rub mineral oil on wooden handles to preserve them. To do this, we simply use a rag and rub oil into the wood until it is saturated. Also in winter, we sharpen our hoes with an aluminum oxide rotary tool sharpener, which fits on a portable drill. Caring for tools in winter is a way to practice quiescence: Like a dormant tree, rest but get ready.

Before we use our harvesting knives, we sharpen them with a 12″ restaurant-grade, oval, diamond knife-sharpening steel. For quick access, we keep the sharpening steel on the same magnets on which we store the knives. We pull the edge of our knives along the steel about five times on each side before use.

To keep our greens harvester sharp, we change its cutting blade every four months and keep extra blades and a spare-parts kit handy.

We charge an extra set of battery packs for each of our battery-powered tools and store these full batteries near where the tools are stored. To preserve battery life, we bring them inside during winter so they are not stored in temperatures below freezing.

A mounted air hose reel is handy to access; it increases the odds that our tires will stay inflated.

The most recent improvements we've made in our tool-care system: We installed a garden hose reel and an air hose reel on our porch, where they are easy to access. We use the garden hose to wash down the Jang seeder, the paperpot transplanter, and other tools after use. The air hose is constantly pressurized from a small air tank on the inside wall opposite the reel. The air hose extends and retracts easily and is outfitted with a pressure nozzle. With this tool in place, our tires now stay full of air.

# CHAPTER 9

# *Less but Better*
# Infrastructure

The theory of "just enough" states that a tiny farm should have just the right amount of infrastructure—not too much, not too little—and that infrastructure should be maximally used. Earlier, I shared inspirations from traditional farm layouts and principles from Dieter Rams that guided us as we designed a just-enough farm. In this chapter, I'll describe in detail what we built and how we built it—and some mistakes we made along the way.

When setting up our farm, we wanted our gardens and buildings to give us the feel of a cottage industry. We wanted a *studio* lifestyle, like a potter or jeweler or cobbler who works from home. We wanted a farm that felt cozy, approachable, and manageable. Through careful design, our gardens and buildings gave us this feel, though it took a few years to settle in after the construction mess.

In a sense, we've spent the past 16 years leaning up our infrastructure. Here is the best way I know of to set up a micro farm. In chapter 10 I offer an even simpler approach at a backyard scale.

## How We Dealt with City Officials

The first step in our building process was to get our plans approved. First, a planning department had to approve our site plan, which had been drafted by a certified engineer. Then, we needed to pass building inspections at different stages of development. These included framing, mechanical, and environmental engineering inspections. It was fun to draw up our dream farm, less fun to get it approved.

To start the permitting process, I first met with the city planner to discuss the overall scope of our project and next steps. She told me we would

Our cedar farm sign.

need plans that included setbacks (the distance between our buildings and neighboring land), the size and location of our buildings, and a utilities plan, along with many small details like the number of parking spaces we'd like and the material used to construct our lane.

Over the next several months, I made many trips to her office, with different iterations of our plans to show her. In the end, our plans were approved, though not without many conversations to clarify our intentions and the city's requirements.

These negotiations can be stressful—there is a lot on the line—and confusing because city codes can be cumbersome to read. But if you plan to build in a city, or anywhere where permits might be required, you will need to work with officials. Here are two tips, based on our experience.

## TIP 1: GET EVERYONE IN THE SAME ROOM.

We made the offer to buy our land contingent on getting a go-ahead on our plans from the city first. We didn't want surprises. Before we even drafted plans, I asked the planning department for an interdepartmental meeting. I said I'd like to meet with everyone in one room who might have a stake in our project—the fire marshal, engineer, building inspector, or anyone else. She told me those meetings were typically reserved for larger projects, like an apartment building or a school. I convinced her that our project was unusual, if small, and that it warranted a frank talk. She agreed to the meeting.

To prepare, I wrote a paper describing our current farm and our vision for a new farm in Goshen, and I gathered pictures of our operation. I handed out the paper and pictures. I told the group, "We're excited to move our farm into Goshen. But we want to know now, before we invest in land and plans: Does this project fit the city's vision? If so, are you willing to work with us?"

Each official considered the project and potential problem points. I answered questions and addressed concerns. By the end, everyone in the room agreed to work with us and to be flexible if needed, so long as we agreed to follow the proper processes.

We ran into a few hiccups as we set up our farm, but I had that meeting to fall back on. For example, a few times, on-site inspectors weren't willing to approve certain parts of our building project that weren't familiar to them. In those cases, I called the department chairs and kindly reminded them of our meeting—and of their commitment to work with us. We were always able to come up with a way forward that met both the city's concerns and our needs.

Luckily, more cities are creating spaces for urban farming. Many officials are excited to work with farmers. Sadly, in some cases they aren't so accommodating because urban farms are so uncommon. The key in either case is open communication and early buy-in.

## TIP 2: STRENGTHEN YOUR BATNA.

BATNA is an acronym that stands for best alternative to a negotiated agreement. The concept was articulated by Roger Fisher and William Ury, of the Harvard Program on Negotiation, in their book *Getting to Yes*.[1] The stronger your BATNA, the better position you are in to negotiate an agreement. Essentially, strengthening your BATNA means always having a strong plan B in mind.

For example, let's say you planned to have granola for breakfast, but your lovely partner ate it all for a midnight snack. You might be tempted to throw a fit. However, upon reflection, if you develop a BATNA, you are prepared with an acceptable solution in an egg and toast.

We've found time and again that if we strengthen our BATNA, we are less stressed and more confident when talking with city officials. Here are a few examples. When submitting plans for our first greenhouse—a 34′ × 148′ structure—we had no idea whether the city would approve it. A greenhouse like this had never before been built inside of Goshen. A few weeks after I

## Good Land for Micro Farms

We used table 9.1 to guide our land search to make sure that the land we chose contained features essential for growing vegetables and making a living.

Searching for a piece of land for a micro farm is a job that requires both a heart test and a head test. The land should *feel* right—it should be aesthetically pleasing and feel like home, especially if you plan to live there. This is the heart test.

Table 9.1 is the *head* test. We used it to check our instincts. In fact, we visited several plots before deciding to buy the land we now farm. A number of them passed our heart test—they were beautiful. But they failed the head test—they were too waterlogged or shady or not close enough to town.

For every plot we visited, we entered a number for each critical feature, from 1 being bad to 4 being perfect. For example, if the plot of land was very close to markets—say, within biking distance— we'd give it a score of 4. If the plot might require cutting down a few trees, we'd score it a 2 and perhaps get a quote for the cost of tree removal. In the end, for a plot to be viable, we said it should score 20 points or more, with no dealbreakers.

We picked 20 as a reasonable minimum number. Maybe yours is higher. Any lower and you'll have significant work to do to remediate the property. This remediation work—cutting trees, installing utilities, improving drainage, and so on— can be rewarding and might save you money on the property purchase price. But before jumping into a purchase, it's best to know beforehand the work and costs that lie ahead to bring a less-than-ideal plot up to a standard for growing.

We took a few other steps before deciding to buy. We sent in a soil sample to test for nutrients and toxins. Nutrient

**Table 9.1.** Tiny Plot Suitability Checklist

| | Points | Cost |
|---|---|---|
| 1 = DEALBREAKER. THERE IS NO REMEDY TO THE PROBLEM. 2 = A REMEDY EXISTS, BUT IT WILL COST US. 3 = FINE. NOT GREAT, BUT IT PASSES. 4 = EXCEPTIONAL. | | |
| 1. Close proximity to markets | | |
| 2. Well-drained soil | | |
| 3. Sun exposure | | |
| 4. Flat terrain | | |
| 5. The right size | | |
| 6. Access to water and electricity | | |
| TOTAL (POINTS MUST BE HIGHER THAN 20) | | |

levels were very low, but that didn't worry us—deep composting restores soil health. Fortunately, no toxins were found. To find a lab for soil testing in your area, look online for private laboratories or contact your local USDA service center.

Finally, we consulted the USDA's soil survey website (https://websoilsurvey .nrcs.usda.gov/app) to understand soil types on the land. The site can also help you read slopes, identify water tables, and determine its suitability for agriculture.

submitted a sketch of the greenhouse, the building inspector called me into his office and grilled me with questions such as, How do you plan to anchor this building? What is the wind rating? Has an engineer signed off on this? We discussed options that would satisfy the city in each of these areas. In the end, the building was approved. But if it hadn't been approved, we had backup plans to propose—such as smaller greenhouses with fewer mechanical systems or simpler hoophouses. Though we never used them, these alternatives, or BATNAs, gave me confidence when talking with the inspector.

Another example: Our plans called for a well, which we would dig, instead of hooking up to costly city water. But before talking over this point with city officials, we'd discussed the situation with a plumbing and excavating contractor. We developed BATNAs—such as digging a pond to pump water—in the event the city wouldn't approve of a well. In the end, we settled on a compromise with the city: We were required to hook up to city water and to use it in our house, but outside of the house, we were permitted to rely on well water.

In both cases, BATNAs gave us leverage to negotiate an agreement that worked in our best interest. Fortunately, city officials were mostly fair-minded and willing to entertain outside-the-box solutions.

The biggest mistake we made at the planning and permitting stage was to underestimate the costs that go along with these planning steps. Site plans, permits, and inspections all came with higher price tags than we anticipated. To do it again, we would have been more diligent getting quotes first.

Another mistake: As we planned our farm lanes, we underestimated the width required for delivery trucks to turn around. We later widened our lanes—after one truck got stuck and another ran into a bed of mint.

We also failed to account for future development. We decided after two seasons to add another greenhouse. This required another round of conversations and approvals that could have been avoided had we included that possibility at the beginning.

# How We Set Up Electric and Solar Systems on the Farm

Most of our farm systems are powered by the sun. We planned carefully to make this happen. Here I'll explain two passive solar systems that we employ, followed by electrical and solar systems we put into place. It's beyond the reach of this book to describe these technologies in detail; rather, this list merely shows a sun-based approach to farm infrastructure that works well for us.

With new incentives to go electric, solar-powered farming is more possible now than ever. I encourage you to seek out grants, low-interest loans, and tax credits geared to making renewable energy more affordable.

Our farm is not 100 percent electric-powered. For a few months in the spring, we heat greenhouses with natural gas to get a jump on tomato season, though we use the heater less each year. We still use a gas-powered delivery vehicle, though we plan to replace our delivery vehicle with an electric option when prices are more affordable. With Swadeshi as our ultimate goal, we hope to someday deliver with bikes, but right now the roads aren't safe enough for bicycle transportation.

## THE SUN-TRAP DESIGN

Like the connected farms of New England, we arranged our buildings to trap the sun's heat on the south side while blocking cold winds from the east, west, and north. We purposely designed our barn-house to be long and skinny to help catch radiant heat. We positioned the house on an east-west axis, 20′ south of a wind-blocking pine grove. Our market garden sits to the south, about 30′ from the barn-house. Our kitchen garden beds lie directly against the south face of the barn-house. A forest blocks winds on the east side of the gardens, and our greenhouses block winds from the west. The result is a U-shaped area that warms quickly in the mornings and stays warm in the evenings, after other areas cool off. Temperatures on the south side of the barn-house are sometimes 10 degrees warmer than on the north. When it is sunny in the winter, snow on the north side of the barn-house will stay on the ground weeks after it's melted on the south side.

We enjoy the benefits of living and farming in a sun trap. In our kitchen garden, garlic and mint are up weeks ahead of usual in the spring, and parsley stays fresh late into the fall. Our kids' sunflower garden thrives there, as well. In the market garden, we've seen little wind damage on our plants compared to other places we've farmed. Because soils warm quickly in the

We arranged our buildings to create a sun trap. The barn-house faces south for passive solar heating. Photo courtesy of Adam Derstine.

spring, we can direct seed in February, weeks earlier than we could start planting at our old farm, which is in the same growing zone. Also, thankfully, row covers and plastic tarps don't blow away as frequently as they have in other places we've farmed.

## PASSIVE HEATING

Passive heating means trapping heat from the sun in your building through strategic window placement, insulation, and thermal mass. Our barn-house is passively heated through a generous amount of window glazing on the south side. On many sunny afternoons in fall, winter, and spring, we can turn off the mini-splits (described on page 167) because a small fire in a wood stove is sufficient to heat the whole barn-house.

We attached our propagation house to the south side of our barn-house, adjacent to our processing room. We built the structure with locally milled tulip poplar lumber and 8 mm polycarbonate sheets and installed a cement floor with a drain in the middle for easy cleanup. The benches are 36″ tall and 36″ deep, and shelves on the south wall provide extra growing space.

A 10′ × 24′ propagation house, attached to our barn-house, provides a controlled environment for starting young plants. Front polycarbonate panels can be removed in summer, as shown. A window connects the propagation house to our processing room, allowing excess heat to be put to use warming interior space.

The room is vented with a 36″ endwall louver and 24″ endwall fan, and heated with a Modine Hot Dawg natural gas unit heater, which we rarely use. In midsummer, we also prop open a 4′ × 4′ window in the roof.

We placed a large casement window between the two rooms. On sunny afternoons in winter, we can open the windows and vent excess heat into our processing room. This regulates temperatures in both rooms and supplies the processing room with warm, moist greenhouse air—the best kind of air in which to work on a dry, cold winter day.

## SOLAR PANELS FOR A SMALL FARM

We invested in enough photovoltaic (PV) panels to power our farm's heating and cooling and most other energy needs, with logistical help from a local solar advocacy organization. To size the system, we estimated the number of kilowatt hours our farm might use in a year and gave this number to our installer. Half of the energy generated powers our living quarters, and the other half powers the farm.

The main electrical draws on our farm include the following, in order by amount of use:

1. Coolers for produce
2. Mini-split cooling and heating systems
3. LED grow lights
4. Fans, tools, and other incidentals

Thirty-four solar panels on the barn-house roof provide sufficient power for business and household use.

I recommend using a load-calculating spreadsheet, available online, to help you determine your total load. An electrician should also be able to help with this task.

Most areas allow you to connect your PV system to the grid. In essence, with a grid-tied system, you are selling the power that you generate to the utility company and then buying it back when you use it. In the best-case scenario, you sell your power for the same price at which you buy it. Fortunately, we are able to do this. With grid-tied systems, the utility company serves as a kind of battery bank. Alternatively, you can install batteries to store your own energy, but these systems cost more and require upkeep, as the batteries need to be maintained and, eventually, replaced.

## HEATING AND COOLING WITH MINI-SPLITS

Mini-splits (also called heat pumps), powered by electricity, are widely considered to be the cheapest and most efficient way to heat and cool indoor spaces. They are better for the environment, using half as much energy as conventional fossil fuel–powered home heating systems. The heat pump compressors sit outside of the house and heating/cooling "mini-split" heads are mounted on the inside.

One mini-split head is dedicated to our processing room. We rely on it to keep the space cool in summer for washing greens and warm in the winter for starting early tomatoes under grow lights.

Heat pumps work by moving heat around rather than generating it through a combustion fuel source. In the summertime, they work like conventional AC units, pulling heat from rooms and dumping it outside, while blowing cooler air back in. In winter, refrigerant moves the other direction, allowing mini-splits to bring heat inside while removing cold air to the outside.

In reality, we rarely use our mini-splits in winter in the living quarters of the barn-house, because we rely on the centrally located woodstove.

## CONSTANT PRESSURE WELL PUMPS

We installed a constant pressure well pump to deliver water pressure more evenly for our crops. These pumps use sensors to dial up and down the amount of pressure applied to water instead of pumping the same amount of pressure every time, resulting in better performance while less electricity is being used.

For example, if just one irrigation sprinkler is calling for water, a low amount of pressure is applied. If ten sprinklers are running, the pump will supply enough pressure for all of them. In our experience, with a constant pressure pump, seeds germinate more evenly because our irrigation coverage is more uniform.

## LED GROW LIGHTS

Another change we made was to convert to LED grow lights, which use a fraction of the amount of electricity required by older light systems. They are also simpler: The diodes never need to be replaced.

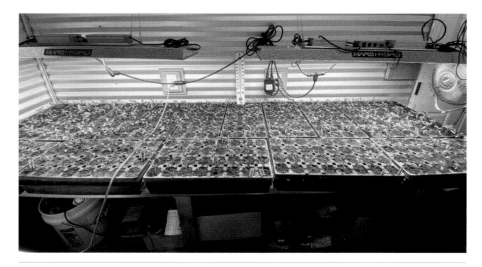

Mars Hydro LED grow lights help our young tomato plants grow.

## Rethinking Refrigeration: How We Deliver Food within Four Hours of Harvest

Dieter Rams said that good design is both innovative and environmentally friendly. We kept these principles in mind as we thought about the problem of micro-farm refrigeration.

At our old farm, we used an 8′ × 8′ walk-in cooler to store our crops. This allowed us to spread out our harvests. But the cooler always bugged us. It used more electricity than other appliances on the farm. Plus, it was awkward to move totes in and out of such a small space.

At our new farm, we decided to eliminate the cooler. Instead, we deliver food as soon as we pick it—often within four hours. To accomplish this, we schedule our harvesting to happen on Tuesday and Thursday mornings, typically in less than two hours' time. After crops are picked, we wash and sort them in the processing room for another few hours. In summer, our full crew of four workers helps harvest and process. Then we back our delivery van up to the door and load it. The processing room is cooled with a mini-split, as stated, thus crops are kept in relatively cool conditions. The van is typically ready to leave by

We cool this three-door refrigerator with an air conditioner and a CoolBot. This uses a fraction of the electricity required by a walk-in cooler.

169

noon, though in peak season we need a few extra hours.

There are special circumstances that do require us to refrigerate, though on a smaller scale. There is not enough time in the morning before a farmers market to harvest all of our crops. So, we harvest many items the night before and store them in a three-door refrigerator, which is much lower cost to operate and more environmentally friendly than a walk-in cooler. We power this refrigerator with an air conditioner controlled by a CoolBot programmer. (CoolBots are widely used on farms to save energy, and if you are not familiar with them, I recommend reading more about them online or in *Growing for Market* articles.) We also sometimes store tomatoes in a tomato cooler (the converted germination chamber) in the event that we overpicked tomatoes and want to keep them for a few days.

There are several benefits, besides saving energy costs, to the approach. We save two moves—walking food into the cooler and picking it up again to remove it from the cooler. There is no "cooler confusion," where totes and labels get mixed up in such a small space. And, of course, chefs love the fast turn-around time.

Loading the delivery van immediately after harvest. Note that the van can back right up to the processing room door. The shed roof allows us to load in the rain and shades the vehicle from the scorching sun in summer.

Specifically, we bought two Mars Hydro TSL 2000 2′ × 4′ light fixtures. We suspended them above a table in our processing room with adjustable rope hangers. In summer, when we're not using the lights, we shorten the ropes and store the lights near the ceiling, freeing up the table to sort tomatoes.

One downside to LEDs is that prolonged exposure to the light can damage your eyes. We never look directly at the diodes when they are powered on, and we wear grow-room glasses when working with plants under the lights.

## A Minimalist Processing Room

The processing room is the nerve center of a tiny farm. Everything flows through it. It should be designed "thoroughly, down to the last detail," as Rams would say. Also, it should be spare, containing only the tools and

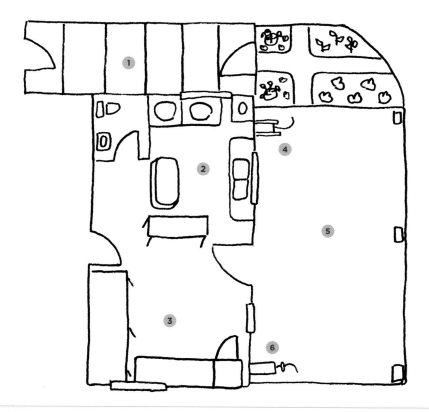

The floor plan of our processing room and hosing station. 1. Attached lean-to propagation house (10′ × 24′). 2. Greens washing station (12′ × 12′). 3. Sorting station, with three-door cooler and germination chamber (12′ × 12′). 4. Water-hose reel. 5. Lean-to vegetable hosing/loading area (15′ × 28′). 6. Air-hose reel.

supplies absolutely required to wash and package food. The room will probably become the busiest spot on your farm. Avoid chaos with a minimalist mindset.

A good rule of thumb when designing vegetable washing spaces is to think in terms of 12′ × 12′ squares—about the size of a typical kitchen. In a square this size, two people can work comfortably. I recommend 3′ walkways between squares—this leaves enough room to comfortably walk while carrying crates and totes. Workers should be able to access each square independently, without having to walk through other workspaces.

Our 15′ × 28′ processing room contains two squares, or workstations: a greens-washing square, for rinsing, spinning, and packaging greens; and a vegetable-sorting square, for grading and packaging larger crops like cucumbers and tomatoes. Adjacent to the processing room is a washing shed—another square, in which we hose down root crops.

## A SMALLER, BETTER GREENS WASHING STATION

When we first set up a greens washing station at our old farm, we thought the best approach was to go big—to leave plenty of space for moving around large totes of fresh-picked greens. But the result was a lot of wasted steps. At our new farm, we compressed.

The first step in greens washing is to dunk greens in cold water, as soon as possible if the leaves are warm, to remove the field heat. The dunking is also an opportunity to clean dirt from the leaves. Many types of tanks will suffice—I've seen farmers use bath tubs and cattle troughs—but restaurant-grade stainless steel is ideal because it sanitizes easily. (To sanitize, we wipe down all surfaces that food might have touched with a rag and then spray with a hydrogen peroxide solution; we let the surface air-dry.) We wash our greens in a two-basin sink that sits under a window overlooking the rest of the farm. Washing greens can be a lonely job, and we wanted the person doing it to feel connected to the rest of the farm, not slinked away in a corner. Some farms use pressurized air to bubble the wash water, which can help tumble the leaves. At our small scale, we opted to simplify and tumble leaves by hand.

Next the greens are moved to a homemade funnel built of plastic suspended over a 2′ × 4′ frame, where they fall into a spinning basket (the basket is drilled with holes every few inches) and placed into a washing machine converted into a greens spinner. We have two such spinners, to speed up the work. To create them, we removed the washing machine tops and moved the control panel to the front. We use the spin cycle to dry the leaves. Two minutes in the spinner typically suffices. After spinning, the greens are dumped into a large tub and placed into clamshells, bags, or totes for customers.

The best way to design vegetable processing spaces is to divide them into 12′ × 12′ workstations. Our processing room includes a greens washing station, shown here, and a vegetable sorting station.

Each of these processes occur within the 12′ × 12′ workstation so that whoever performs the job only has to turn around to move from one step to the next. The table where greens are bagged can be accessed from two sides so multiple people can help with this step and talk comfortably while they work. This is an example of design with people in mind.

## A SIMPLE SORTING STATION

Much of our work, post-harvest, involves grading food—pulling out "seconds" so that customers receive the most blemish-free products possible. At our new farm, we wanted a dedicated, well-lit station for this task.

We grade most large crops on a sturdy 30″ × 8′ stainless steel table that sits steps away from our produce refrigerator and from the door leading to the delivery vehicle. Bags and boxes for packages are stored on a shelf below or on a rack an arm's-length away.

When we have a high volume of tomatoes and cucumbers to process, we pull the table away from the wall so it can be accessed from both sides, or we set up additional folding tables to expand the sorting area. This modularity allows us to quickly adjust our grading setup as needed throughout the season.

## THE HOSING STATION

Outside the main door to our processing room, we built a shed roof to cover a cement pad, also 15′ × 28′. The roofing metal and posts for the space were salvaged from the corn crib that we demolished. The cement slopes away from the building. Just off the edge, we dug a French drain to collect water. Without the drain, that area would quickly turn into a mud puddle. The drain consists of a perforated 4″ PVC pipe buried 6″ in the ground and covered with river rocks.

On one corner, attached to the barn-house, we mounted the water-hose reel for washing crops. On the other corner, we mounted the retractable air-hose reel.

A hanging hose reel keeps our hosing station tidy.

Using chains, we hung tables—repurposed stainless-steel shelves—from the roof as a surface for washing food. We hang the tables to make it quicker to clean underneath them (there are no legs to sweep around). We store harvesting crates and totes under the roof, along with our golf cart.

Again, the space is multifunctional. We use it to host workshops of up to 25 people (if it's cold, we move workshops inside to the processing room). The area is at the center of the farm, with a view of our fields and greenhouses, making it an ideal spot for events. We like that the space,

We installed a drain off the edge of the porch to carry water away from the building. This perforated pipe will be covered with river rocks.

once again, puts people at the center of the farm, where they belong.

## IDEAS FOR DURABILITY AND FLOW IN THE PROCESSING ROOM

We decided to finish the processing room on the interior as we did the exterior—with Tyvek house wrap covered by galvalume sheet metal—so that we can hose down the space with a pressure washer for occasional deep cleaning, keeping in mind the 5S principle (see page 78) to start cleaning at the corners.

The main door between the room and the outside is 42″ wide—wider than standard doors—to make it easier to move through while carrying vegetables and for wheelchair access. The room is lit by high-lumen LED fixtures so we can easily see our work. The mini-split heats and cools the space. The walls are 6″ thick and sheathed with another 1″ of rigid insulation, making it low-cost to condition. As stated earlier, the attached propagation house also helps to heat the room. There are two drain basins in the room, one in the middle and one under the sink; the cement slab floor slopes to these basins.

We don't have a storeroom per se. Instead, we keep most supplies in the workstations where we use them, and we keep them in low inventory to save space. We store towels and wash rags on a shelf near where they are used. We keep tomato boxes on a wall a few steps from the tomato sorting table. We tuck clamshells and plastic bags underneath the packing tables where those supplies will be used. Likewise, supplies for starting seeds can be found under the grow-light table, near the germination chamber/tomato

## Pipes Underground

Many farm books skip over utilities, but utilities can't be glossed over. They undergird every aspect of production.

If you plan to farm land close to your house, it might be tempting to assume water, gas, and electric systems in the house can serve the farm, too. Sadly, that's not usually the case, even for tiny-scale farming. As with other farm infrastructure, you want just enough—not too much or too little. Utilities are expensive to install, and there is no good reason to put in what you won't use.

The IOWA Y1 water hydrant is a durable design with a larger size casing than standard hydrants. We buried water lines 4' deep to avoid freezing. We have four hydrants on our ⅓-acre farm—one in each greenhouse and two in the field—so that no growing bed is more than 100' from a hydrant.

### Water Lines

The first calculation you will probably need to make on a new farm is how much water pressure you need. To determine the size of our well and the diameter of our underground water lines, we told our contractor the maximum amount of water we planned to use at one time in peak season. Specifically, we told him the maximum number of gallons per minute (GPM) our irrigation devices require and the pressure rating (psi) of the devices.

The maximum amount of water we planned to use was 59 GPM. The maximum psi was 30. We also told the contractor how many linear feet of water lines we expected to put in, which came to 500'. The contractor put these three numbers into a formula and determined that we would need the following:

Well: 6" diameter
Pump: 3 HP
Water pipe (underground): 1½"

This configuration has served our ⅓-acre farm well. I recommend consulting a technician to determine your own needs.

Installation of water lines and a well is a job best left to excavators and plumbers. These high-pressure systems require an experienced hand to assemble, and a contractor will be familiar with local codes that apply.

### Gas Lines

We use Modine natural gas unit heaters in our propagation house for early springtime starts and in greenhouses to

**Table 9.2.** GPM at Clay Bottom Farm

| Irrigation Device | Number of Devices | GPM per Each | Total GPM | Max PSI Rating |
|---|---|---|---|---|
| Xcel-Wobblers | 10 | 3 | 30 | 20 |
| Dan Modular hanging assembly | 40 | 0.5 | 20 | 30 |
| Drip tape | 2,000′ | 4.5 per 1,000′ | 9 | 15 |
| **MAXIMUM GPM USED AT ONE TIME DURING PEAK SEASON: 59** | | | | |

We irrigate with farm-built risers. Parts: Senninger Xcel-Wobbler head, 4′ length of ½″ steel pipe, Orbit step-in spike base. Each riser irrigates an area about 40′ × 40′. Photo courtesy of Caleb Mast Photography.

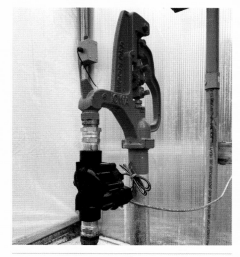

Our hydrants are controlled by solenoid valves, connected to the Orisha Automation system (see page 180). This allows us to precisely control irrigation, with timed settings, from an app on our phone or desktop computer.

protect early tomatoes in late March and April. To save costs, we have stopped heating greenhouses in the winter months unless we need to thaw crops for a morning harvest.

To determine gas line size for a greenhouse, use a greenhouse BTU calculator, available at https://www.gothicarch greenhouses.com/Greenhouse-Heater-Calculator.htm. Simply input your greenhouse surface area, desired temperature, and your location. The calculator will determine what size heater you need. Give this information to your plumber or excavator, and they will determine the line size required.

cooler. By storing supplies under tables, we maximize use of a small space while keeping workstations clutter-free.

One regret with the design of our processing room is that we did not finish the cement floors with a sealant. The floors always look dirty, even after mopping them. We plan to coat the floors with a more durable sealant in the future, but this would have been easier to do before we moved in.

We also regret the placement of our outlets. Most of them are near the floor, as in a house, and breakers trip about once per month  when water splashes up on them. Higher outlets would have been better.

# Better Greenhouses

When we downsized our farm, we shrank our greenhouse space nearly in half because we wanted less greenhouse space to manage. Instead of four green-houses, we put up two. Yet we didn't want to sacrifice productivity. We wanted to do less but better, boosting productivity in our greenhouses with less work.

One of our greenhouses is 34′ × 148′ and the other is smaller, measuring 21′ × 75′, though they both include the features in this section. I partnered with CT Greenhouse and Orisha Automation to design a kit with these features, called the Tomato Powerhouse, which can be found at https://ctgreenhouse.com/powerhouse.

## PEAK VENTS: THE BEST NATURAL VENTILATION FOR A GREENHOUSE

I've come to believe that all greenhouses, even in the north, should be built with a peak ventilation system, especially as the climate continues to change, causing more heat waves. Peak ventilation means creating a way for heat to escape through the top of your greenhouse. Heat rises, so the peak is the best place to let it out. Simply put, the best way to increase productivity in a greenhouse is to ventilate properly.

Here is how peak ventilation works thermodynamically. When a sensor in the middle of the greenhouse reads temperatures above a set tempera-ture—say, 80°F (27°C)—at least one roll-up curtain (on the side of the greenhouse) and the peak vent open simultaneously via motors. Cool air from near the ground is pulled up through the opening created by the roll-up curtain and mixes with air inside. The hottest air at the top escapes. The greenhouse becomes a chimney, causing natural cross-ventilation.

In midsummer, we add a shade cloth to the top of the greenhouse, which decreases heat buildup further and provides dappled shade, making for a pleasant environment for tomato pruning. In the summer, we sometimes prefer working in the greenhouse rather than outside.

At our new farm, we chose to install rack-and-pinion peak ventilation systems. This type of vent looks like a chicken wing flapping open and closed at the top of the greenhouse. The vent spans the length of the greenhouse. It is lifted up by racks—small aluminum posts—held in place by pinions, or round gears. The pinions are secured to a rolling bar that runs the length of the greenhouse. When the bar rolls, the gears activate and the roof opens. A thermostatically controlled DC, low-voltage motor powers the roll bar.

Peak vent systems are common in Europe, but they are less common on small farms in the United States, in part because of their complexity and cost. Until recently, affordable, simple peak vent systems, designed appropriately for small farms, did not exist in the US. That's starting to change. A relatively low-cost rack-and-pinion peak vent kit for smaller greenhouses can now be purchased from CVS Supply, an Amish-owned greenhouse manufacturer in Ohio, reachable by phone at 1-877-790-8269. They also sell a retrofit version than can be added to an existing greenhouse. CT Greenhouse also supplies a kit.

A note about installation: I've installed two peak vent systems, and my experience is that their assembly is no more complicated than building the greenhouse itself. If you have basic construction skills, it is within your ability to do it. For me, it took as long to assemble this system as it took to build the rest of the greenhouse. So, assume that it will double your construction time, but it is worth the effort.

A peak ventilation system is open to let out hot air.

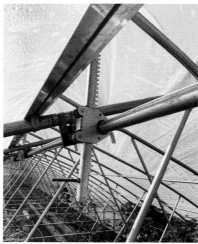

Peak ventilation framing members, before and after plastic installation. This peak vent is opened via a rack and pinion and controlled with a DC motor.

If you don't have mechanical experience, I recommend working with an electrician to hook up the motor and control.

## AFFORDABLE AUTOMATION FOR MARKET GARDENERS

Greenhouse automation means that a computerized programmer controls heating, cooling, irrigation, and other mechanical systems in a greenhouse. For decades, large commercial greenhouses have relied on automation, but until recently these systems were too costly and complicated for market gardeners. As with peak vents, however, new affordable and simpler kits are now available.

We use two types of automation systems on our micro farm. First, we employ the VCU2-24 controller from Advancing Alternatives, which can be found at www.advancingalternatives.com. This unit can roll up and let down sidewall curtains and peak vents that are equipped with a motor, and it can act as a thermostat for heaters. Use push buttons to set your desired activation temperature for each device. We hung a temperature sensor in the middle of the greenhouse, connected to the control panel via a wire. Motors for each device are also connected to the control panel via wires.

Recently, we updated the controller to a Wi-Fi-enabled app-based system from Orisha Automation. This controller can be purchased at www .products.orisha.io. I receive notifications on my phone from Orisha's app if temperatures are outside of a set range. I set desired temperatures and

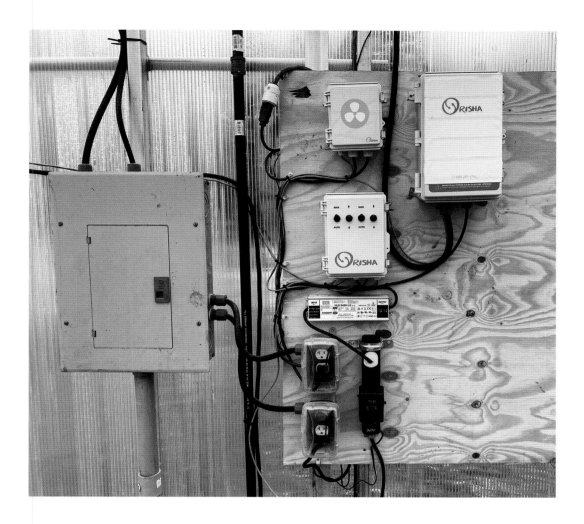

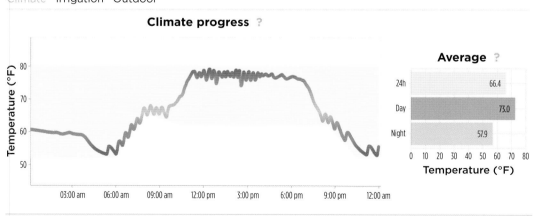

This app-based system from Orisha Automation puts our environmental and irrigation controls on one page and allows us to view historical data.

This DC motor from Advancing Alternatives controls a side curtain.

monitor historical temperature trends in the greenhouse from the desktop or phone app. The system can also activate vents in the case of high humidity levels. The app controls our heaters and, through the use of solenoid valves, all of the farm's water hydrants. In essence, the Orisha system puts all of our environmental and irrigation controls on one page. I can adjust conditions with a few clicks.

In the winter, we turn off power to the side curtain motor and ventilate the greenhouse exclusively with the peak vent. In this way, no cold air is allowed to rush across our plants, an especially important consideration for young tomato plants.

Starting in mid-April, when we require more ventilation than the peak vent alone can supply, we open the switch to our automated side curtain again, allowing it to operate in tandem with the peak vent. Once the weather stabilizes, around the end of May, we manually open the other side curtain for maximum airflow. We will leave this side curtain open all summer.

With Orisha, we tell the software program our desired average morning, daytime, and nighttime temperatures—in sunny and in cloudy conditions—and it uses an algorithm to control vents and heaters in the most efficient way possible to maintain those averages and to avoid the heater kicking on while a vent is open.

While these systems required an upfront cost and time to install, they quickly paid off. They save time because I am no longer running out to the greenhouse to manually ventilate every time the sun appears from behind a cloud; we now have the flexibility to leave the farm for days at a time. Also, plants grow faster because they bask in ideal photosynthesizing temperatures. With automation, our productivity and profits have increased with less work.

## FIVE MORE TIPS FOR GREENHOUSE PRODUCTIVITY

Here are five more features that we put into our newest greenhouses to increase their productivity:

## The Ultra-Low Tunnel: A Low-Cost Starter Option

We use this ultra-low tunnel to get an early start on spring crops. It is basically a large box with cables strung every 5′ to support a piece of row cover. The cables are held tight with turnbuckles and are propped up with short wooden posts.

The box is covered by a large piece of Agribon AG-50 row cover, held tight with ball bungies and leftover shade cloth clips. To prevent ripping, we applied ZIP System Flashing Tape where the clips grip the fabric. The design could easily work on smaller garden boxes, even on 4′ × 8′ garden beds.

1. *Tall posts.* We asked our manufacturer to deliver extra-tall posts to give our greenhouse sides 2′ of additional height. Standard greenhouse kits supply posts that stick up 4′ out of the ground. Ours are aboveground about 6′ and inground about 3′. The added cost was minimal, but the advantages are significant. With taller posts, we can trellis tall crops on the growing beds along the edges of the greenhouse, and we can traverse in and out of the tunnels from the sides.

## Shading a Greenhouse

A shade cloth pulled across the top of a greenhouse will decrease the amount of UV light that penetrates through the greenhouse plastic, thus cooling the air inside.

We pull shade cloth over our greenhouses in the first week of July, and we remove it in late August, for a total covering period of six to eight weeks. This helps protect plants and fruits from direct overhead sun, and it cools greenhouses during the hottest part of the season. In other growing zones, these dates will vary.

*Tip #1:* Use shade cloth that leaves the sides of your greenhouse exposed. Our sheets cover just the top two-thirds of our greenhouses. This way, direct light can still reach your plants in the mornings and evenings.

*Tip #2:* Use shade cloth clips and ropes, tied to attachment points on the greenhouse, to pin down shade cloths. Unlike a sheet of plastic, a shade cloth, with thousands of small holes, is not easily blown by the wind. On our 148′-long greenhouse, we attach the shade cloth at just eight points.

*Tip #3:* To pull a shade cloth over a tall greenhouse, tie ropes along the edge every 30′ to 50′. To prevent the shade cloth from ripping, tie the rope around a ball bunched up in the shade cloth. Using tape, attach a softball to the other end of each rope. Next, toss the softballs over the greenhouse, walk to the other side, and pull the shade cloth up and over.

Shade cloth helps control the temperature in our greenhouses. Note that we only cover the peak to shade midday sun.

2. *Sidewalks.* Greenhouse entryways are high-traffic areas that can quickly turn into muddy messes. To avoid the problem, we installed sidewalks along the endwalls of our greenhouses. On our big greenhouse, there is a sidewalk both inside and outside. This allows for a small working and storage area inside the greenhouse.

To construct these sidewalks, we built a frame with leftover 2″ × 6″ lumber and packed the floor underneath with a vibratory plate soil compactor, before a cement truck poured the concrete. If your land is poorly

Installing tall posts on our second greenhouse. Note the use of mason line to ensure the posts are plumb. We also used a level on each post when pounding them in.

Preparing to pour concrete sidewalks on the endwalls of a greenhouse. Note the lumber separating the inside and outside sidewalks. An undivided cement slab could conduct cold into the greenhouse.

We used salvaged cement chunks to construct this sidewalk in front of a greenhouse endwall.

drained, I recommend adding a few inches of sand and tamping it with a compactor before pouring cement.

3. *Bifold doors.* We have tried many greenhouse door systems, and our favorite by far is the bifold door. This type of door was common on 19th-century carriage houses, and they are still made by Amish manufacturers. The door is very simple. It uses no electricity and few parts. It opens manually, with the help of a pulley and weight. It hinges in the middle and opens outside of the greenhouse, so there is no interference with plants or trellising systems inside. The door, when opened, also provides a handy roof in the event of rain. We purchased our doors through Silvercraft in Middlebury, Indiana, which can be reached at 574-825-8757.

On greenhouses narrower than 24′, to save costs, I recommend building your own doors with locally purchased lumber or steel or aluminum tubing. The doors on our smaller greenhouse are framed with 2″ × 2″ lumber. They are mounted with hinges that swing out so we can grow plants right up to the endwall.

A bifold door makes an ideal greenhouse door. The design is simple and effective. There are no electric parts, the door takes up no room on the inside of the greenhouse, and it functions like a small roof overhang when opened.

4. *Steel baseboards.* On our new greenhouses, we installed steel hat purlins as baseboards. Hat purlins are used for roof and wall support in steel construction. They make an excellent low-cost baseboard for greenhouses. Typical baseboards, made from wood, rot quickly. These will last a lifetime. We purchased our hat purlins (as well as hoop-house frames and many other hoophouse parts) through CVS Supply in Ohio, which can be reached at 877-790-8269. These parts can also be purchased through Nolt's Greenhouse Supplies.

Hat purlins make an ideal baseboard system.

5. *Swedish skirt.* A Swedish skirt is a ring of insulation, laid right on the ground, around a greenhouse. We graded soil up to the edge of our greenhouse so water falling off the roof drains away from the structure. Then we cut 4′ × 8′ pieces of 1″ rigid foam insulation in half and laid the 2′ × 8′ pieces of insulation along the sides. Next, we covered the insulation with landscape fabric and installed steel landscape edging along the outer edge. We pinned the landscape fabric to the hat purlin baseboards using leftover aluminum extrusions from the greenhouse.

The primary advantage of the skirt is that growing beds along the edge of the greenhouse do not freeze during the winter, allowing us to grow crops in them. This system also permanently keeps weeds from creeping in underneath the baseboards—a nasty problem on many farms. Finally, the insulation acts like an eave on a house, keeping water away from the building.

A Swedish skirt involves laying rigid insulation horizontally along the sides of a greenhouse. This is a simple way to insulate the greenhouse, stop weeds, and shed water. Note that we graded soil away from the greenhouse.

Here I am adding 18″ of vertical insulation under a greenhouse endwall.

On the endwalls, we insulated by installing 18″-wide rigid insulation vertically, underneath framing members. We don't use the Swedish skirt method along endwalls because we walk in this area.

Adding insulation around a greenhouse pays off quickly. Insulation lowers heating bills and makes the structure more functional. It is another example of using a smarter approach to increase the productivity of a small space—of building better not bigger.

# CHAPTER 10

# Getting Started: A Plan for Selling $20,000 from Your Backyard

Here is a plan for a backyard micro-enterprise using just 5,000 square feet with minimal startup costs. The preceding chapters showed how we used the get-small approach to set up a farm using ⅓ acre—about 15,000 square feet—for production. However, for many readers—those in the first year of growing or those wanting to farm part-time—the smaller plan presented here might be more realistic. Simple is best, and this is the simplest way I know of to start farming.

With a waste-free mindset, a 5,000-square-foot plot can yield $20,000 worth of produce, with an allowance for some inevitable losses. The plot size in this plan is 50′ × 100′, divided into 12 growing beds. To see if you have enough room, you can pace your backyard or plot. Measure your stride. Mine is about 3′, so for me the plot size is 16 steps in one direction and 32 steps in the other. Of course, you can adjust the size of the plot to fit the room you have.

If desired, you can easily expand this plan. For example, if your goal is to sell $40,000 worth of produce, simply double the number of beds or double their length. (Of course, you'll need to sell twice as much, too.)

To fulfill the plan, set aside about six months to set up infrastructure and to prepare your plot, ideally starting in the fall before your first season. The total start-up costs (including for seeds, tools, compost, and the starter hoophouse) in this plan are about $7,700, assuming you buy everything new. This price can be significantly lowered, possibly cut by more than half, with a bit of salvaging and reusing. Tending the crops will require two to three days per week during peak season.

## Han-nō, Han-X ("Half-Agriculture, Half-X")

Farming need not be full-time. In Japan, there is a popular trend to take up part-time farming. The movement, called *han-nō, han-X* ("half-agriculture, half-X"), is based on the belief that connecting agriculture with another job or pursuit can offer a sense of rootedness through a connection to the seasons. The movement has inspired many Japanese to move from urban to rural areas, and the government has even begun to incorporate the movement within its national agriculture policy.

A recent article in the *Japan Times* featured as examples a professional *djembe* (a West African drum) player who grows an ancient red rice that he sells at his concerts, and inn owners who grow vegetables for the inn and invite their guests to join them in harvesting wild tea leaves.

The term *han-nō, han-X* was popularized by author Naoki Shiomi, who wrote about the idea on a website and in a book published in 2003. Shiomi took inspiration from an earlier book by environmentalist Jun Hoshikawa, former executive director of Greenpeace Japan, who said,

> *Although* han-nō, han-X *is not a panacea for positive social change, it is a sure route to a fairer and more sustainable world. By inviting people to try growing food while earning a modest income, the magic is that you are likely to realize you don't need as much money as you thought you did. The agricultural lifestyle brings not only a fair amount of food, but also happiness and fulfillment for free— which you had previously chased by spending money.*[1]

The plan I present is designed specifically for those wanting to try the half-agriculture, half-X model.

The model I provide here is one path to joining the micro farm revolution, accessible to just about anyone—or to a group of friends—with a decent-size backyard, access to water, and the necessary start-up capital. This chapter would make a good business plan for those wanting to farm part-time (see "Han-nō, Han-X," above), for recent high school or college graduates, or as part of a gap year between studies. Good work is gratifying, and microscale farming could be a good transitional activity.

The "Five-Step Quick-Start Guide" (see page 195) is a minimalist approach to growing the four crops in the plan. In fact, this is the approach to crop production that we use on our farm, only our beds are slightly longer (75′) and our greenhouses are bigger. See Clay Bottom Farm's Instagram feed (@claybottomfarm) for real-time photos and videos throughout the seasons demonstrating these and more techniques. We also offer workshops and an online course in market gardening geared toward microscale growers (see claybottomfarm.com), and there are many other online and in-person resources available as well.

## A Garden Map for Earning $20,000 from 5,000 Square Feet

This map calls for a focus on four high-margin crops: tomatoes, salad greens (lettuce and Asian greens), spinach, and kale. Of course, you can add more crops, and in appendix A, I provide minimalist growing tips for additional crops. I'd start with these four, however, because they are in high demand

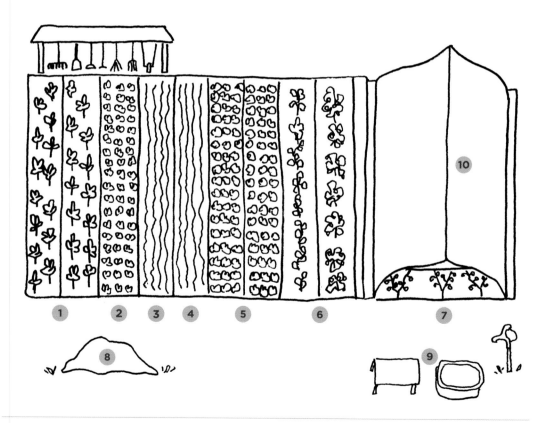

A simple farm plan to earn $20,000 from a 5,000 square foot plot. 1. Kale from transplants (two beds), on landscape fabric. 2. Direct-seeded spinach. 3. Direct-seeded lettuce. 4. Direct-seeded Asian greens. 5. Head lettuce for salad mix (two beds). 6. Slicing tomatoes (two beds). 7. Heritage tomatoes (three beds). 8. Compost. 9. Vegetable washing station. 10. Starter hoophouse..

**Table 10.1.** Income Plan

| Crop | Total # of beds | Income per bed | Total income |
|---|---|---|---|
| Heritage tomatoes | 3 | $1,800 | $5,400 |
| Slicing tomatoes** | 2 | $1,050 | $2,100 |
| Salad greens | 8* | $1,200 | $9,600 |
| Spinach | 2* | $1,000 | $2,000 |
| Kale | 2 | $450 | $900 |
| **TOTAL INCOME** | | | **$20,000** |

*Double-cropped, meaning that you would have four beds of salad greens and one bed of spinach in the spring, and then these beds would be flipped mid-season to produce a fall crop.

** You can substitute with cherry tomatoes, if you like. I describe growing techniques for both types in Step 5 below.

across the country—an easy sell. They are also high-margin crops because they are multiharvest vegetables: plant once, harvest lots of times.

This garden contains twelve 3′ × 50′ beds, separated by 12″-wide pathways. Note that the salad greens and spinach beds are double-cropped—with a spring planting followed by a fall planting. Three of the beds—for heritage tomatoes—are covered with a small hoophouse. (For bonus income, the hoophouse could be used to grow greens in the winter.)

This plan assumes average sales of $1,000 per week for 20 weeks—May through October—to achieve $20,000 in gross sales. In reality, sales will be lower in spring and fall than in midsummer, but I've made an effort here,

## The 4-2 Farm Start Method

It is better to gain confidence doing a small thing well than to overwhelm yourself with more than you can handle. In your first season, I recommend a 4-2 approach: Focus on four crops and sell to two types of customers. The four high-demand crops in this plan are a great starting point. Your customers might include one restaurant and a farmers market, or a co-op and a school cafeteria. (I don't recommend operating a CSA in the first year because of the complex planning required.)

Starting simple is best. Plants don't mature overnight; likewise, farming takes time to master. Schumacher, late in life, implored his audiences "to start from where you are and grow from there, rather than build up a vast structure which ends as an empty shell or hot air balloon of words."[2] I couldn't agree more.

**Table 10.2.** Prices for Four Crops

| | Wholesale price/pound | Retail price/pound | Retail price/unit |
|---|---|---|---|
| Heritage Tomatoes | $5 | $6.50 | $7/quart |
| Slicing Tomatoes | $3 | $4.50 | $5/quart |
| Salad greens | $11 | $14.00 | $5/clamshell (5 ounces) |
| Spinach | $11 | $14.00 | $5/clamshell (5 ounces) |
| Kale | $8 | $10.00 | $4/bunch of 10 leaves |

through succession planting, to spread sales and work out as evenly as possible, to avoid midsummer burnout (see table 10.4).

The backbone of the plan is salad greens—selling an average of 40 to 50 pounds per week for all 20 weeks. The other staple is tomatoes. Kale and spinach make up the remaining income. The dollar values per bed are based on data we've collected here at Clay Bottom Farm. Prices will surely be higher in some areas of the country and lower in others.

*A note about setting prices:* Most farming guides ask you to perform complex analyses—where you track labor, supplies, seeds, and marketing costs per crop—to determine your prices. There is a shortcut. Lean says that it is not up to you to determine prices—it is up to the marketplace, to customers. My favorite quote from Taiichi Ohno at Toyota: "Costs don't exist to be calculated—costs exist to be reduced." The fast way to determine prices is to take a walk around a nearby organic food store—where produce prices are likely highest—and use them to help you determine prices. If you are selling wholesale, then ask your buyer what they pay other wholesale suppliers. You want your prices to match the high end of the current marketplace.

Table 10.2 shows our current prices for the four crops in this plan.

# A Five-Step Quick-Start Guide

The appeal of micro farming is that it has a relatively low barrier to entry. Really, there are five main steps you need to take if you decide to take the leap.

## STEP 1: ACQUIRE TOOLS AND SUPPLIES

Before you start growing, you'll want to assemble the tools and supplies listed in table 10.3. This starter kit, based on my 16 years of growing experience, will set you up for success. See chapter 8 for pictures and descriptions

## Budget for a 5,000 Square Foot Start-Up Market Garden

In table 10.3 are all the items you need to get started. Of course, you can save costs by buying used or using what you already have to replace some of these items.

These figures are projections based on data from our farm. Your first year could be a boom! Or a bust! Expect either. But if you have setbacks, don't get discouraged. Learn from them, make changes, and keep going.

A note about refrigeration: In addition to the supplies in chapter 8, you will need refrigeration, unless you live close to your markets and can deliver upon harvest. To store the greens, you will need two upright refrigerators or a large chest freezer that you power down using the INKBIRD that I describe on page 147. You can also use a nonworking two- or three-door commercial refrigerator that you cool with an air conditioner and CoolBot, as shown on page 169.

of many of the tools. The supplies are referred to in more depth in step 5. Support your local businesses and buy as many items as you can from small local hardware and farm supply stores close to home. Otherwise, search online.

When setting up, remember the principle *just enough*. You want the right number of tools and supplies—not too much, not too little. I've carefully curated this list to supply the essential items you need for a lean micro farm start-up.

Another first step is to build a tool rack. This plan calls for just a handful of tools, so you won't need a shed dedicated to tools. The tool rack can be as simple as two posts with a beam, about 5′ high, attached to the posts. An optional small roof will shed rain and prolong the life of the tools. Alternatively, hang the tools on an

You don't need to buy a shed for tools. A simple tool rack will keep tools accessible.

**Table 10.3.** Starter Kit Tools and Supplies

| | Item | Cost (approximate) $ |
|---|---|---|
| Seven vital field tools | Bed preparation rake from Johnny's Selected Seeds | 100 |
| | Wheel hoe with Hoss fixed-blade sweeps | 310 |
| | Aluminum scoop shovel | 50 |
| | Adjustable-width rake | 20 |
| | DeWit half moon hoe | 80 |
| | Narrow collinear hoe from Johnny's Selected Seeds | 50 |
| | Clarington Forge garden fork | 100 |
| Harvesting tools | Curved grape shears | 20 |
| | 6" stainless steel restaurant-grade produce knife | 20 |
| | Harvest bucket from Johnny's Selected Seeds | 32 |
| | 18-gallon Rubbermaid totes × 6 | 140 |
| Planting tools | Jang seeder and 3 rollers[A] | 600 |
| | Trowel | 20 |
| | 72" white oak tomato stakes and twine from Nolt's Produce Supplies | 100 |
| General supplies | 5 oz landscape fabric 6' × 250' | 150 |
| | 20,000 BTU weed burner torch for landscape fabric | 40 |
| Irrigation supplies | ¾" × 100' garden hose | 100 |
| | Parts for two Xcel-Wobbler head riser assemblies[B] | 100 |
| Washing station supplies | 50-gallon Rubbermaid stock tank for washing greens | 140 |
| | Choice Prep 5-gallon greens spinner | 100 |
| | Packaging (bags or clamshells—as per the customer) | 100 |
| Seed-starting supplies | Potting mix (local)[C] | 50 |
| | Plug flats[D] | 50 |
| | Mars Hydro TSL 2000 grow light[E] | 270 |
| | INKBIRD for germination chamber[F] | 30 |
| | Used slow cooker for germination chamber | 10 |
| Incidentals | Watering can, office supplies, plant labels, tomato hooks, etc. | 200 |
| Starter hoophouse | See step 3 (page 198) for ideas | 2,300 |
| Seeds | | 500 |
| Compost | 62 yards @ $30/yard[G] | 1,860 |
| **TOTAL START-UP COSTS:** | | **$7,642** |
| **TOTAL GROSS INCOME PROJECTED:** | | **$20,000** |
| **NET INCOME:** | | **$12,358** |

A To save costs, you could seed by hand (or use the lower-cost Earthway seeder) for the first season, but I advise buying a Jang seeder as soon as you can afford it. Jang roller sizes that I recommend: F-12, X-24, LJ-24. See Table 8.1.

B I discuss a Xcel-Wobbler head set-up on page 177.

C See page 203.

D 32-cell, 72-cell, and 128-cell—buy what you need depending on crops grown, per table 10.4.

E This grow light is shown on page 168.

F See "A Simpler Germination Chamber" on page 147.

G Prices vary widely—shop around.

outside wall of your house or outbuilding, under an eave. The point is to store the tools outside, near the garden, where you can grab them quickly.

The supplies required (row covers, stakes, etc.) will be few in number and can fit inside of the hoophouse or in your garage, if you prefer.

For a processing station, all you need to begin are a tub to wash greens, a greens spinner, and a hosing table. Set these things up in your garage so you are out of the sun. Otherwise, outside under a tree, pop-up tent, or lean-to will suffice for your first season.

## STEP 2: BUILD A DEEP-MULCH PLOT

To work up the ground, kill the existing turf with a tarp or large sheet of opaque plastic, pinned down with sandbags (cinder blocks could work here, too) for at least two weeks, and then till or loosen the ground with the digging fork. Tillers are available at rental centers. (If you are in a hurry, rent a sod remover from your local rental center and pull off the sod by hand.) Then follow the approach outlined in chapter 6 to add 4″ of good compost to the surface and to lay out the beds. For a 5,000-square-foot plot, you will need 62 cubic yards of compost.

## STEP 3: BUILD A STARTER HOOPHOUSE

A small starter hoophouse is essential to protect tomatoes and to grow greens in winter, if you choose to do so. Place the hoophouse in the most easily accessible spot in your plot, but leave 3 to 4′ around the perimeter for the Swedish skirt (see page 188) and drainage.

I mentioned in chapter 9 that, at Clay Bottom Farm, we now build greenhouses with peak vents, tall sides, and automated controls. Once you have experience with hoophouses, this is the style I recommend. If you are a novice grower and still learning construction skills, start simpler, with a low-cost starter hoophouse.

Below are a few options, among many available, to consider.

- Quonset Greenhouse from Nolt's Greenhouse Supplies: 12′ × 48′ for about $1,365
- Gothic Caterpillar Tunnel from Farmers Friend: 16′ × 50′ for about $2,360
- "1000 Series Hoophouse Frame" from Greenhouse Megastore: 16′ × 48′ for about $1,699 (does not include plastic or baseboard)
- "DIY Greenhouse Kit" from Bootstrap Farmer: 12′ × 60′ for about $1,075 (kit optionally includes a bender for you to bend your own pipes)

These photos show how Ben Matthews from Warfleigh Cottage Garden in Indianapolis built his urban farm plot with a deep-mulch approach. He covered the ground with a tarp, removed it and loosened the ground, then added compost. His garden is abundant. Photos courtesy of Ben Matthews.

199

Heritage tomatoes, which are grown in the hoophouse, will require at least 5′ between rows. Thus, a 12′ hoophouse will support two beds of tomatoes; a 16′ hoophouse will support three. If you are in the north and expect a lot of snow, I recommend a Gothic-style hoophouse, which has a bit of a peak. Otherwise, a Quonset shape will suffice.

At this stage, I recommend shaping the ground around your hoophouse so that water drains away. The Swedish skirt will also improve drainage.

## STEP 4: MAKE A GARDEN PLAN USING VALUE SHEETS

Your tools and supplies are in place. You've worked up your plot and added compost. You've built a hoophouse. It's time for a specific plan.

Remember, start with the customer and work backward from there. Use a Swadeshi mindset: Sell hyperlocally. As I explained in chapter 5, I recommend taking a small field trip. Create a map of the food buyers located closest to your farm. Offer them these four crops (plus others if you'd like). Ask what they want, when they want it, and how much, and create your value sheets. *This is the most important day or two you will spend on your farm business.* Remember to essentialize: Choose a "vital few" restaurants/grocery stores/markets to focus on—those who seem most committed to supporting you with consistent orders—rather than selling to everyone you visited.

Then use the value sheets to adjust the number of beds for each crop, according to the feedback you received. Don't worry for now about planting precisely the right amount down to the last inch—just fill the beds with crops roughly proportional to their popularity.

Hold your plan lightly—you want to be ready to change plans at a moment's notice to keep up with changing demand from your customers. It's easy to spend too much time farming with pencil and paper, projecting yields and trying to predict how every week of the season will go. It's better to start where it counts—with the customer—and to stay nimble. Also, in your first season, you will be learning many skills; it's unlikely everything will go as you plan.

## STEP 5: GROW!

The crop-growing methods I describe in the remainder of this chapter are specifically suited to a micro farm. As you grow your skills, you can also easily expand. Simply lengthen your beds—from 50′ to 75′ or 100′, or add more beds. (We use 75′-long beds to accommodate the paperpot tools, as I explain on page 234.)

# Sell to Your Neighborhood Through a Roadside Stand

You might be surprised at how many potential customers live right around the block or right next door.

To test your neighborhood market, I suggest sending an email through your neighborhood association, if one exists. Or simply set your food out on a card table with a large inviting sign and see what happens. If sales pick up, add a tent, or buy or build a small roadside stand.

Simplify pricing. I suggest selling with a uniform price structure: for example, $4 per item or three items for $10, like Tiny Giant Farm. If you can't

staff the booth, consider using a mountable cash dropbox with a slot for customers to self-pay. As you build a network of relationships, perhaps start a neighborhood CSA, where neighbors prepay and pick up food once a week.

Other ways to draw interest from neighbors include offering free food—extra tomatoes, zucchini, watermelon, and the like when you have a bumper crop—or hosting a meal. There is no better way to weave in the local community than through food, the ultimate conversation starter.

**Table 10.4.** Seeding Calendar for a $20,000 Garden

| Crops | | Direct-seed (DS) or transplant (TP) | Plug flat size* | Date to seed (in ground or in flats)** |
|---|---|---|---|---|
| Tomatoes | Heritage tomatoes | TP | 32-cell | 3/15 |
| | Slicing tomatoes | TP | 32-cell | 4/1 |
| Salad greens | Baby lettuce mix and Asian greens | DS | | Bed #1: 3/1, 7/20 |
| | | | | Bed #2: 4/1, 8/10 |
| | Multileaf lettuce | TP | 128 | Bed #1: 4/1, 6/20 |
| | | | | Bed #2: 4/15, 7/1 |
| Spinach | | DS | | 3/1, 8/25 |
| Kale | | TP | 72 | 4/1 |

*Plug flats start under grow lights indoors and are ready to transplant in four to five weeks.

**These dates are based on our experience in northern Indiana, zone 5b. Please adjust them for your own area. I recommend using the seed starting date calendar on the Johnny's Selected Seeds website.

Here, I offer a doable, minimalist approach to high-productivity growing for folks with little or no gardening or farming experience. The methods are specifically tailored for tiny plot production. Of course, this template need not be followed to the letter. Perhaps you want rows that curve, or maybe you want to focus on bell peppers, garlic, and sunflowers. There is every reason to adjust these plans to fit your own needs. The information presented here is merely a suggestion based on what has worked well on our farm.

Go on then, make a farm.

# A Minimalist Method for Heritage Tomatoes

## Heritage Tomato Quick-Start Basics

*Varieties:* Margold, Marnouar
*Starting method:* Place 1 seed per cell into 32-cell plug flats
*Optimum germination temperature:* 80°F (27°C)
*Spacing:* 5′ to 7′ between rows, 1′ apart in-row spacing, about 50 plants per bed
*Trellising method:* Qlipr system or Rollerhooks

On many tiny farms, heritage tomatoes are the highest-profit-margin crop. They can be a challenging crop to grow, though with these techniques, it is possible to grow them successfully in your first season. We grow Margold, a yellow tomato, and Marnouar, a purple tomato. The chefs we sell to tell us that together, Margold and Marnouar make their plates "pop" with color. They sell fast at farmers markets and to grocery stores as well.

Margold and Marnouar tomatoes, our go-to heritage tomatoes.

# Heirloom Versus Heritage Tomatoes

Heirloom tomatoes are traditional seed varieties whose seeds have been passed down for many generations. Heritage tomatoes, by contrast, are hybrids; their seeds cannot be reused.

Heritage tomatoes are bred to look and taste like heirloom tomatoes, but with more vigor and disease resistance. Heirloom tomatoes offer lower yields and a higher cost of production than heritage tomatoes. I recommend heritage tomatoes in this plan because of their economic reliability. That said, there are good reasons that others prefer to grow heirlooms—including their better flavor and their independence from corporate ownership. Additionally, growing heirlooms supports ongoing genetic diversity.

## STARTING SEEDS

The simplest way to start tomatoes is to use 32-cell plug flats and to transplant the tomatoes when they are relatively young—about four weeks old. This avoids the need to "pot up," or move tomatoes to larger containers. We buy our plug flats through Nolt's Produce Supplies.

Use a quality potting mix, such as Vermont Compost Company's Fort Lite. When filling your trays, make sure the mix is slightly moist: when you squeeze a fistful of potting mix, water should drip out. Then press one seed into the middle of each cell and cover with a bit more potting mix. Next, water the trays (we use a Wonder Waterer wand) and put them into a germination chamber (see page 147) set to 80°F (27°C). Check on them daily until the seeds sprout. Most varieties require three to five days.

The trays must be pulled from the chamber within 12 hours of germination and set under grow lights. (See a photo of a simple grow light setup on page 168.) We set the lights on timers so they are on between 7 a.m. and 7 p.m., and we keep the lights about 18″ above the plants, which leaves room for a watering can. We prefer a simple gooseneck watering can for the job. It puts water at the base of the plant without wetting the leaves.

Tomato starts in 32-cell plug flats, ready for transplanting.

## A Word about Potting Mixes

In keeping with our deep-mulch approach, we prefer compost-based potting mixes. Many companies sell quality compost-based mixes, and we have certainly not tried them all. However, Fort Lite from Vermont Compost Company outperforms other potting mixes that we've used.

Tip: Fort Lite is biologically active upon arrival. It relies on microorganisms in the potting mix to make nutrients available to seedlings. Thus, it is critical to keep the potting mix moist, even when you are not using it; without adequate water, microbial activity grinds to a halt.

For stout plants, let nighttime temperatures dip to as low as 60°F (16°C), but ensure that daytime temperatures hover between 70 and 75°F (21–24°C) and not above 80°F (27°C). A good place for this setup might be your living room near a heater or in the basement.

### TRANSPLANTING TIPS

We grow all tomatoes in landscape fabric. The fabric simplifies fieldwork because it holds water in the ground and stops the weeds. The fabric also prevents tomatoes that have fallen to the ground from coming into contact with the soil, which quickly leads to rot. When fallen tomatoes are left to decompose in place, their seeds can germinate and become an annoying weed in the winter greenhouse and the following spring.

The day before transplanting, we water the growing beds well with drip tapes. A few hours before transplanting, we give the seedlings a good bottom watering soak (we set them in cafeteria trays or something similar, filled with water). With wet soil and wet transplants, there is usually no need for additional watering on transplant day. Drip tape is simple to set up; just follow the manufacturer's instructions. We buy our drip tape supplies online from IrrigationKing. I recommend 1″ or 1½″ main line and 8 mm-thick tapes that have emitters spaced every 4″.

### TRELLISING AND PRUNING HERITAGE TOMATOES

About a month after tomatoes are transplanted, prune the branches and suckers off the bottom 12″ of stem to increase airflow. The bottom leaves are

Transplanting tomatoes into landscape fabric. To avoid transplant shock, set in plants in the late afternoon or on a cloudy day.

Lay down four runs of drip tape per tomato bed to ensure that the entire root zone stays moist. Our drip tapes have emitters spaced every 4″. On sunny days, in our soil conditions, we typically water for about 20 minutes each at noon, 2 p.m., 4 p.m., and 6 p.m. Less watering is required on cloudy days. Tomatoes will frustratingly crack if watered too heavily at fruiting time. If you see a lot of cracks, try watering less.

## Simplify Fieldwork with Landscape Fabric

We use landscape fabric under widely spaced crops, like tomatoes, that will be in the ground for several months at a time. Landscape fabric allows rainwater and oxygen through while blocking light, thus killing weeds like a tarp.

Using a 20,000 BTU weed torch and plywood template to burn 5″ holes into landscape fabric for transplanting. We have reused some pieces of landscape fabric for more than 10 seasons.

It is much faster to lay down the fabric than to battle weeds all season. Landscape fabric is superior to organic mulches, like straw or wood chips, because it is quicker to apply and lower-cost. It is also more effective: Unless straw or chips are laid down very thick, weeds will find a way through.

Another reason to use landscape fabric: to create a spacing template. There is no need for measuring tools when transplanting in landscape fabric. We use 6′-wide 5-ounce fabric from DeWitt.

*Tip #1:* Mid-season, use a battery-powered leaf blower to clean dirt and plant debris off the tarps. We do this once per week in our tomato rows. This makes it easier to clean the tarps later on, and it helps maintain sanitary conditions and

Pinning down fabric with 8″ sod staples. Landscape fabric is an essential supply on a tiny farm.

Spraying landscape fabric with a solution of OxiDate to sanitize it.

tidiness. Wear a mask to avoid breathing in the dust.

*Tip #2:* After removing tarps at the end of the season, give them a hearty shake and broom or hose off any debris left on them. This will prevent a moldy mess when you pull them out again in the spring. We store our tarps, folded up, in a corner of a greenhouse.

*Tip #3:* Before using them the following spring, sanitize tarps to prevent the spread of disease. We use the Jacto PJB battery backpack sprayer filled with a solution of OxiDate, an organic peroxide-based spray designed to control plant pathogens. We spray both sides of each piece of fabric, typically on the day we install them.

always most prone to disease; they will need to be removed sooner or later, and it's much faster to remove them now when they are small than when they are four times larger.

As the plants grow, pinch or prune off "suckers"—the fruiting branches that grow between stems and the two plant leaders—every week. If you procrastinate even one week, then your work can be four times more burdensome because there will be exponentially more vegetative matter to drag out of the greenhouse.

If you find that you have waited too long to prune and face an arduous task, do not let this dampen your spirit. Farming can be pleasurable because it is good work. Be thankful that you have more good work to do. Next time, prune the plants on time.

Tomatoes are trained to two leaders per plant.

When plants are knee-high, after about four weeks in the ground, you can start to trellis them, picking two leaders to train up twine. We use two approaches: Rollerhooks and the Qlipr system. We started with Rollerhooks 16 years ago, and now we are switching to Qlipr hooks gradually as twine runs out on our Rollerhooks. With either system, hooks are hung two per plant, spaced 12″ apart from each other in the row, on ⅛″ 1 × 19 stainless steel cables strung across the greenhouse and tightened with ratchet strainers.

Trellising like this is common, and I recommend watching videos about the

Leaning and lowering tomato plants. We use a mega-lean approach, where we slide the growing tips up to 8′ in one go once the plants reach the cross-ties.

practice—or, better yet, visiting an experienced grower—to get a sense for what's involved.

## A NOTE ABOUT GROWING CONDITIONS

Tomatoes are a warm-season crop that likes temperate weather, with no direct rain on their leaves. Ideally, soil temperatures at transplant time are 60°F (16°C) or higher. Some growers lay down clear plastic for a few weeks ahead of planting tomatoes to warm the soil.

A word of caution: Cold air rushing across young plants causes them stress. If you *must* roll up side curtains to vent heat, consider blocking the plants from the wind by placing 2′-wide insulation, or some other rigid panel, along the bottom of the opening.

## MANAGING PESTS AND DISEASES

To prevent tomato hornworms, we use DiPel DF, an organic insecticide made from *Bacillus thuringiensis*, a soil-based bacterium that makes proteins that are toxic to leaf-eating caterpillars. We spray it on all of our plants with the Jacto sprayer as soon as we spot our first hornworm. Typically, one or two sprays per season are all that is required.

Another pest problem: aphids. These small insects multiply rapidly, and by the time you see them, it might be too late to control them. We use a preventative approach. We order green lacewing eggs from ARBICO Organics early in the season and set up biweekly deliveries of additional eggs until August. Lacewing eggs are the most effective low-cost biological control for aphids that we have encountered. The eggs take three to five days to hatch upon arrival, so if you have an immediate infestation, a better solution is to first spray with PyGanic Specialty, an organic insecticide.

## FERTILIZING TOMATOES

With the deep-mulch system, we have phased out the use of mineral supplements for tomatoes. Compost alone is sufficient to supply tomatoes what

Note that the lower leaves of these plants are well pruned. This allows for maximum air flow and minimizes disease.

they need. The one exception: If it appears that the tomatoes aren't fully ripening, toward the end of the season, supply liquid potassium sulfate to the plants. Tomatoes pull a lot of potassium from the soil, and by late season, a greenhouse full of tomatoes will sometimes suffer from potassium deficiency. Overall, however, there is no need to complicate things with an arduous fertility plan. Trust the compost.

## A WORD ABOUT GRAFTED TOMATOES

Grafting tomatoes involves splicing together a hearty rootstock with a scion—your desired tomato variety. Grafting can increase plant vigor and yield and decrease disease, but it requires considerable skill. It also increases costs. While we grafted tomato plants for many seasons, we no longer do so. With a deep-mulch approach, grafting is not necessary. Also, new tomato genetics are more robust, diminishing the benefits of grafted plants.

## HARVESTING

Once they ripen, I suggest harvesting tomatoes at least three times per week. In most cases, you'll want to harvest tomatoes at 90 percent ripeness—when they show good color but before they start to soften. This ensures they will

Harvesting tomatoes with the help of a cart.

keep well for customers, and they will continue to ripen after harvest. When we harvest for a food co-op, we harvest at 70 to 80 percent ripeness, at the request of the co-op, to prolong shelf life further.

I recommend using over-the-shoulder harvest buckets, available at Johnny's Selected Seeds, to collect tomatoes. Use the curved grape shears to cut off each tomato, leaving a bit of stem. If you remove the stems from heritage tomatoes, the tomatoes could rot where the stem was pulled off. Use care when placing tomatoes in the buckets so that stems don't puncture other tomatoes. Mid-season, when there might be hundreds of pounds to harvest in a go, a flatbed garden cart is a welcome aid.

## SELLING

We sell wholesale tomatoes in 10-pound, single-layer tomato boxes. If we need to store them, we use our germination chamber, which doubles as a tomato cooler in summer. We set the thermostat for the chamber to 55°F (13°C). Any colder and the tomatoes will be mealy when they're removed. At warmer temperatures, they can ripen too quickly.

The key to selling all the tomatoes you grow is to develop a market for all grades of tomatoes. We sell our firsts (blemish-free tomatoes) to high-end restaurants and our seconds to chefs who plan to cut the fruits into pieces—they don't mind small cracks. We sell thirds—canning tomatoes—through an email list we've developed over the years, to folks who enjoy home canning but can't pay retail prices for organic produce.

# Slicing and Cherry Tomatoes

For both slicing and cherry tomatoes, we use the same methods as with heritage tomatoes for starting seeds, transplanting, irrigation, and pest and disease management.

Clementine and Mountain Magic tomatoes.

The only major difference in our growing technique is with trellising and pruning. For slicing and cherry tomatoes, we use a much simpler staking method.

Here's how it works. For slicing tomatoes, we start by pounding in 72″-tall, 1″-thick oak tomato stakes between every two tomato plants. On the ends and every 12′ or so in the middle of each row, we also pound in 7′ tall metal T-posts. The posts on the ends are angled outward by 20 degrees. As plants grow, we weave twine between them, and the twine is also twisted around each wood and metal post, in a "stake-and-weave" pattern. The metal posts support the tomatoes once they are big. More than once, tomatoes with only wood stakes for support have tragically fallen over.

## Slicing Tomato and Cherry Tomato Quick-Start Basics

*Slicing varieties:* BHN 589, BHN 871

*Cherry varieties*: Clementine, Mountain Magic

*Starting method*: Place 1 seed per cell into 32-cell plug flats

*Optimum germination temperature*: 80°F (27°C)

*Spacing*: 2′ between rows; slicing tomatoes 18″ apart in-row spacing, about 33 plants per bed; cherry tomatoes 24″ apart in-row spacing, about 25 plants per bed

*Trellising method*: Stake-and-weave

Cherry tomatoes grown with a tall stake-and-weave method. Note the 10' EMT conduit between each plant. We sandwich the plants with twine, stretched horizontally on both sides of each plant, every 18". To promote airflow, we prune to four growing tips between each conduit every time we apply twine. This is a fast and effective way to trellis cherry tomatoes.

Harvesting cherry tomatoes by hand. In a field setting, just let the tops grow back to the ground once they reach the tops of your EMT stakes.

This picture shows two trellising systems. On the left is a standard stake-and-weave system for slicing tomatoes. On the right is a tall stake-and-weave system for cherry tomatoes. For slicing tomatoes, we place one stake between every two plants. For cherry tomatoes, we place one stake between each plant. With both systems, we use twine to create a net that sandwiches the plants as they grow.

Once, in the middle of a lean growers workshop, the participants heard a crash, and we all went to see a row of staked tomatoes lying flat on the ground. Things happen.

For cherry tomatoes, we use the same practice, but with 10′-tall T-posts placed between each plant. In a greenhouse, we use 10′-tall EMT conduit between each plant. To thin the cherry tomatoes, allow just four tips to grow between each stake (prune off the rest) each time you box them in with twine.

Like all tomatoes, slicing and cherry tomatoes grow best in the protected environment of a hoophouse. In your first season, however, I recommend trying them outside, to save costs. You can always add hoophouses in the future.

## Simpler Ways to Grow Salad Greens

Our salad mix is really a blend of Asian greens, spinach (see page 220), baby lettuce, and leaves from head lettuce, in different ratios depending on the season. The varieties we grow are not fixed. Lettuce genetics improve every season, and we are always sampling new cultivars. But I recommend a minimalist approach—choose a small number of varieties to streamline your work. Even two or three varieties mixed together can create a colorful salad.

Muir is the most heat-tolerant lettuce available.

## DIRECT-SEEDING FOR
## BABY LETTUCE AND ASIAN GREENS

If the weather permits, we start seeding outdoors as early as March 1 for Asian greens (Mizuna and Red Kingdom) and three weeks later for lettuce (Allstar Gourmet Lettuce Mix). We can start selling these fast-growing crops while we wait for the slower-growing head lettuces to mature. We grow the preblended mix of lettuces in order to simplify direct seeding.

## Baby Lettuce
## Quick-Start Basics

*Varieties*: Allstar Gourmet Lettuce Mix from Johnny's Selected Seeds
*Starting method:* Direct-seed using the Jang seeder with roller F-12 and the sprocket set at 13 front/11 rear. Hand-seed by rubbing seeds between your index finger and thumb.
*Optimum germination temperature:* 72°F (22°C). Germinates well at lower temperatures.
*Spacing:* 5″ between rows in a bed. For hand-seeding, space seeds roughly $1/16$″ to $1/8$″ apart (no need to be precise).

## Asian Greens
## Quick-Start Basics

*Varieties:* Mizuna, Red Kingdom.
*Starting method:* Direct-seed using the Jang seeder with roller X-24 and the sprocket set at 13 front/11 rear. Hand-seed by rubbing seeds between your index finger and thumb.
*Optimum germination temperature:* 80°F (27°C). Germinates well at lower temperatures.
*Spacing:* 5″ between rows in a bed. For hand-seeding, space seeds roughly $1/16$″ to $1/8$″ apart (no need to be precise).

To prepare beds for spring mix, we take special care to rake beds flat, using the bed preparation rake, because undulation in the beds will make it difficult to cut greens at a low uniform height, especially if we are using the greens harvester. (In your first season, you can save costs by hand cutting with the produce knife, but get a greens harvester as soon as you can afford to.)

To simplify work, we don't till or fork the soil ahead of planting salad greens. These are shallow-rooted crops; they don't require loose soil.

It is imperative that beds be weed-free. If you suspect a bed is thick with weed seeds, flush them out first by leaving a silage tarp on the bed for at least three weeks before seeding. Some growers use flame weeders to scorch tiny weeds, but we haven't found that to be necessary with the deep-mulch method.

To keep rows straight, we fix our eyes on a point in the distance and walk in a straight line. This is a faster approach than using strings and stakes to guide us. Or use the bed preparation rake with row markers to make a guide. We seed most rows 5″ apart, water the seeds in well, and cover with two or three layers of row cover (to hold in moisture) until the seeds pop.

After seeds germinate, little care is required. We typically do no cultivating; the leaves will soon crowd out any weeds. If you do need to weed, I recommend using the collinear hoe. With the deep-mulch system, no fertilizing is required.

We irrigate as needed using timers and Xcel-Wobbler head risers. In the spring, little or no irrigating is required.

*Tip #1:* Soil for direct-seeded salad greens can be relatively firm. Unless the beds were stepped on when wet and became compacted, we do not loosen soil before seeding.

## Head Lettuce Quick-Start Basics

*Varieties:* Muir, Intercut, Rubygo (for salad mix, pelleted seeds)

*Starting method:* Use 128-cell plug flats or 6″ paperpot chains

*Optimum germination temperature:* 72°F (22°C). Germinates well at lower temperatures.

*Spacing:* 10″ between rows, 4 rows in a 42″ wide bed; 9″ apart in-row spacing, about 67 plants per row; or 6″ with the paperpot method, about 100 plants per row.

Spring mix is a high-demand crop that can be grown at a low cost. Photo courtesy of Adam Derstine.

215

*Tip #2:* Irrigate beds before and after direct seeding for optimum germination. Beds should be evenly moist but not saturated. We use a Dramm 1000PL water breaker nozzle for a gentle spray pattern with even coverage at seeding time.

## FIVE TIPS FOR GROWING HEAD LETTUCE IN SUMMER

We prefer to grow heat-tolerant multileaf head lettuces in summer and into fall rather than direct-seeded salad greens, because their leaves are crisper. They also offer a longer harvest window—sometimes up to a month. As the seeding chart (table 10.4) indicates, we start seeding these in paperpot chains on April 1, but we typically won't harvest them until June. They can be grown equally well by hand-transplanting. We don't typically sell head lettuce as heads. Rather, we trim their tops as we harvest, adding the leaves to our spring mix.

Head lettuce is a cool-season crop, but with extra care, it *can* be grown in summer. Here are five tips.

*Tip #1:* Choose heat-tolerant varieties. The varieties I list above are sweet and crisp in the heat when other greens start to turn bitter.

*Tip #2:* Use a cold chamber for germination. In fact, I recommend using your germination chamber (see page 147) for lettuce—but cool it, using the INKBIRD thermostat to control the compressor, keeping temperatures below 70°F (21°C). Alternatively, start seeds on wet cement, on a shady porch, if hot. If the weather is below 70°F, you can use the germination chamber with a slow cooker.

Summer lettuce seedlings on wet cement under a covered porch.

Once they've germinated, seedlings need continued cool weather, ideally under 75°F (24°C), to grow well. We keep ours cool at a low cost by growing them in different places around the farm, depending on the season. In April, we turn on our heaters in the propagation greenhouse and grow them there. By late May, we pull a shade cloth over the starts in the propagation house to keep them cooler. In summer, we grow them on wet cement on our spray station porch or inside a conditioned room under grow lights.

In summer, we cover new head lettuce seedlings for three weeks with shade cloth. The shade cloth is supported by #9 wire bent with 90-degree angles. In this way, the sides are straight and the shade cloth does not touch the plants.

*Tip #3:* Water the beds very well before transplanting, and avoid overhead irrigating on transplant day to avoid stress on the young plants.

If using the paperpot transplanter, I suggest removing the front plates. This avoids covering the often-floppy transplants, which can be tucked in later by hand.

*Tip #4:* Immediately after transplanting head lettuces in summer, cover with a 50 percent shade cloth held 18″ above the greens with wire hoops. Bend the hoops at 90-degree angles so that the shade cloth does not touch the leaves. With curved hoops, shade cloth will touch lettuces along the edges and burn it. Use two layers of shade cloth when temperatures are above 85°F (29°C). Remove the shade cloth three weeks after transplanting. By then, transplants will be set in well.

*Tip #5:* Mist beds on afternoons when the temperature exceeds 80°F (27°C). We use our Xcel-Wobbler heads, with high-flow (lavender) inserts, set on timers to irrigate for 20 minutes at noon, 2 p.m., 4 p.m., and 6 p.m. We do this misting even if beds are covered with shade cloth.

## TIPS FOR WINTER SALAD GREENS

For extra income, fill your hoophouse with direct-seeded baby Asian greens and spinach immediately after removing tomatoes in the fall, before October 15 if possible. Or start spinach (see page 220) in paperpot trays three

Greens in a high tunnel. Rather than heat our greenhouse all winter to grow lettuce, we grow cold-hardy crops.

weeks ahead of when you expect to remove tomatoes, and set in transplants the day you remove tomatoes. (We've found that it is not worth the hassle to transplant baby Asian greens.) We have no trouble selling every leaf that we can grow in winter. In our experience, lettuces do not perform as well in midwinter as Asian greens and spinach. Thus, we do not often grow lettuce in winter hoophouses in order to maximize yields.

For winter growing, use the same techniques that I describe above for the rest of the year. There is no need to heat the hoophouse in winter—the varieties I recommend grow well without heat and also without row covers. However, supplemental heat and row covers do speed growth.

We pause seeding in December in our tunnels because of the short day length. Typically, winter salad greens are preceded by tomatoes or cucumbers. To prepare beds for winter growing, we cut these plants off as low as possible, remove the landscape fabric, and then rake the beds clean, with no tilling or fertilizing, except for adding 1″ of compost if needed.

## A TRICK TO ACHIEVING MULTIPLE HARVESTS

With proper technique, salad greens can be cut multiple times. From some beds, we've achieved more than six harvests if the weather allows it, though three or four cuttings are more typical. Hence, we call these "cut-and-come-again" crops. We harvest greens into 18-gallon Rubbermaid totes.

The trick to achieving multiple harvests is to cut the leaves no closer than about 2″ from the ground (or a bit higher for head lettuces). In fact, we think of greens harvesting to be similar to trimming a hedge—we take just what we need from the top. For example, if mizuna grows tall and bushy, we will cut just the top 4″. This promotes regrowth and minimizes the amount of stems in your salad mix. This is a case where less is definitely more.

Always harvest when the leaves are dry and cool. If the leaves are warm and wet when picked, they won't store well. We typically harvest early in the morning or late in the evening. In summer, when morning dew can be heavy, an evening harvest will produce the longest-keeping lettuce.

Send the cleanest greens possible to the processing room. We look over our greens and remove weeds and ugly leaves while picking, in addition to inspecting them in the wash-pack station.

## WASHING AND PACKING

After harvesting, move the greens to a cool spot as soon as possible. Never let them sit in direct sun. When it's hot, we keep fresh-cut greens in a refrigerator until they are washed.

We prefer a minimalist setup for washing: Two big basins and a screen are all that's required. Some growers use pressurized air to "bubble" greens, but we haven't found that to be necessary at our scale. We simply dunk about 6 pounds of greens into a sink, press them down and swirl them with our hands, then move them to a funnel, using a homemade screen. The screen is a fish net with 1" net spacing, suspended on a PVC frame. We drilled holes into the frame so that it sinks. The funnel directs the greens into perforated tubs (all-purpose tubs from Nolt's Produce Supplies or Dubois Agrinovation work well), and the tubs then go into a modified washing machine set to permanent spin cycle (search YouTube for videos showing how to do this conversion). Alternatively, spin greens with the Choice Prep 5-gallon greens spinner.

## SELLING

Salad greens are an easy sell to many restaurants. To make your greens stand out in the marketplace, brand them as local. Shipped-in greens can be at a disadvantage because they have already traveled for days; thus, they lack freshness compared to local greens.

Grocery stores are often eager for locally grown greens, but they require special packaging. We sell greens to the local co-op in clamshells that hold 5 ounces of greens, in cases of 20 clamshells. While the clamshells are recyclable and made from recycled

We sell greens in 5-ounce clamshells and in bulk through a local co-op. We sell our salad greens—and almost all other produce—in sanitized totes, trading empty totes for full totes at each delivery, rather than using disposable plastic bags or waxed boxes. This way restaurants and farms can work together to practice Swadeshi, reducing dependence on shipped-in supplies.

ingredients, we still don't like to use so much plastic. We've worked with the co-op to promote an alternative—selling greens in bulk, out of a plexiglass container, to reduce waste. Customers who bring their own containers buy these greens at a lower cost. You can also sell greens in lower-cost plastic bags.

# Spinach: Queen of the Cold

While other crops languish and decline as the weather grows colder, spinach perks up and takes charge. It doesn't fear cold weather—it begs for it.

We grow the variety Seaside, an upright dense cultivar with leaves that stay small, perfect for adding to our spring mix. Seaside is also somewhat heat tolerant, thus slower to bolt in the spring. However, every season we experiment with new varieties on the market.

Spinach is a versatile crop. We sell it separately and as an ingredient in our salad mix. In winter, up to one-half of the salad mix is sometimes composed of spinach (the other half is Asian greens).

## Spinach Quick-Start Basics

*Variety:* Seaside

*Starting method:* For transplanting, place 4 to 6 seeds per cell in 6″ paper chains. Or hand-seed. Or direct-seed using the Jang seeder with roller LJ-24, brush set about ⅛″ above the roller, and sprockets at 13 front/11 rear. We only direct-seed when the weather is below 68°F (20°C).

*Optimum germination temperature:* 68°F (20°C). Germinates well at lower temperatures.

*Spacing:* For transplanting, 8″ to 10″ between rows, 6″ apart in-row spacing. For hand-seeding, space seeds roughly ⅛″ to ¼″ apart (no need to be precise). For direct-seeding, 6″ between rows.

## STARTING SEEDS

The most challenging step when growing spinach is germination. Soils must be cool and moist for the entire germination period. The optimum temperature for spinach germination is 68°F (20°C). It will germinate well if temperatures are lower but not higher. We avoid growing spinach in midsummer. For best results, water the beds before and after seeding. You want to really soak the seeds. After seeding, cover with three layers of Agribon AG-50 row cover to hold the moisture in.

When temperatures are warm, direct-seeding spinach is a gamble. For best results in warmer weather, soak seeds overnight in a jar of water. In the

I keep notes about sprocket and brush settings for the Jang seeder right on the seeder itself.

The trick to good spinach germination is to seed into moist soil and to hand-water after seeding. Then, cover the bed with three layers of row cover to hold the water in.

morning, run the seeds through a sieve and dry on a towel just until they are dry enough to run through the seeder or handle by hand without sticking together.

Another warm-weather spinach trick is to transplant it using 128-cell plug flats with 4 to 6 seeds per cell or with a paperpotter. We start late-summer spinach in paperpot chains. In our experience, transplanted spinach grows as quickly and with the same high quality as direct-seeded spinach, though it does not grow as densely.

When growing from transplants, use the germination chamber as described for lettuce (see page 147). Germination requires about a week. We've had our best results with impervious bottom-watering trays, purchased from Paperpot

We transplanted this spinach using the paperpot transplanter and 6″ paper chains with 4 to 6 seeds per cell.

Company, underneath the paperpot trays, to ensure that the paperpot chains stay moist.

When we transplant spinach with the paperpot tool, we space rows as close together as the tools will allow—from 8″ to 10″ apart. When direct-seeding, our rows are 6″ apart.

*Tip:* Beware of chickweed, which thrives in the same cool weather as spinach. Even a small amount of the weed can quickly take over a spinach crop because it can sprawl underneath spinach leaves. We avoid planting in areas where we've seen dense chickweed in the past, and if we see it, we cultivate it early and often. Always collect chickweed as you cultivate it and remove it from the garden, otherwise it will reroot.

## HARVESTING

Harvest spinach by hand or with a greens harvester, using the techniques I described for lettuce (see page 218). The greens harvester does slice through leaves on occasion, but our customers don't seem to mind that imperfection. Like other baby greens, we cut multiple times. Spinach is washed using the same method as other greens.

*Tip:* Spinach waterlogs easily. Don't let it sit in the wash water for too long. It also bruises easily. Use a gentle touch when moving it during washing.

## SELLING

We sell spinach in 5-ounce clamshell containers to our local co-op, and we sell as much as we can produce. We sell it to restaurants by the pound in reusable totes.

A bed of direct-seeded spinach.

We built this temporary shade house from EMT conduit, cables, and a large piece of shade cloth to protect greens during a few weeks of extreme heat. Tall T-posts would also work well as structural supports for temporary shade.

# A Simple Method for Longer-Season Kale

## Kale Quick-Start Basics

*Variety:* Winterbor

*Starting method:* Place 1 seed per cell into 50-cell plug flats

*Optimum germination temperature:* 85°F (29°C)

*Spacing:* 24″ between rows, about 2 rows per bed. 14″ apart in-row spacing, about 43 plants per row.

*Intercropping:* Optionally, intercrop radishes, bok choy, or head lettuce

Kale is a versatile and always-in-demand crop that can stretch across seasons with the right methods.

### PREPARING THE BEDS

Rake beds smooth, then lay down two runs of drip tape 24″ apart. Next, cover the beds with a 6′-wide piece of landscape fabric into which you've burned two rows of holes for transplanting (see page 206). If you would like to intercrop with the kale, then burn three rows of holes, using the middle row for radishes, head lettuce, bok choy, or other quick spring crops.

### STARTING SEEDS

Start seeds in 50-cell plug flats, and put them in a germination chamber set to 85°F (29°C). As soon as the seeds pop, pull the flats and place them under grow lights.

A new technique that we use on our new farm: We keep kale trays in Giant Plus Garden Trays, available from Greenhouse Megastore. These solid-bottom trays from England are designed to hold four flats, leaving a bit of room on the end for adding water with a hose. We use the trays to bottom-water our kale starts (and other starts as well). We fill the trays with ½″ of water. Once the flats soak it up, we add another ½″. It can take one afternoon or several days for the kale to soak up ½″ of water, depending on the amount of sun and the temperature. Be careful not to waterlog your starts, however. Don't let them sit

A productive bed of kale. Kale thrives with the deep-compost method.

in water for more than a few hours. Bottom-watering is another way to simplify work.

After about six weeks, the kale plants are ready for transplanting. We make sure, on transplant day, to water the growing beds well and to soak the starts. Wet soil plus wet transplants will offer the surest path to success. As with head lettuce, cover with a shade cloth for a few weeks if transplanting in summer.

## INTERCROPPING WITH KALE

We typically transplant head lettuce and bok choy or direct-seed radishes between rows of kale on the same day that we transplant the kale. These "understory" crops are typically ready to harvest just as the kale leaves start to shade them. This can be an advantage, keeping these crops cool in late June and early July.

Kale intercropped with head lettuce.

### EXTENDING THE KALE HARVEST

For long-season kale, we harvest the lowest ring of leaves by hand each week rather than picking from the top, and we rake under the plants to keep the area clean and disease-free. Kale loves nitrogen—it thrives with deep compost. If you have room in the hoophouse, grow early spring or fall kale there.

In the summer, we spray kale about twice each week with DiPel DF—an organic product made from *Bacillus thuringiensis*—to ward off green cabbage loopers. If infestations are thick, then I recommend spraying three days in a row to really interrupt the hatching cycle of the eggs.

We harvest into bunches of about 10 leaves per bunch. To speed up harvest, we put rubber bands around

Kale can be harvested dozens of times in a single season. The key is to harvest from the bottom up.

Kale in a greenhouse. This kale is being shaded by cucumbers.

our fingers and pull them off as we go. If selling by the pound, we pick straight into a tote.

## Preparing for the Next Season

After you've harvested all your crops, I recommend doing three things.

1. *Cover your growing beds with another 1″ of compost.* Do this in the fall, after all crops are harvested. Don't wait until spring to do this work. *Farm like a tree: Work ahead.* Spring is always a rush to get plants in the ground. Your second season will run more smoothly if you prepare your beds ahead of time.

Remember the rule: Leave roots in the ground. Cut kale stalks and tomato stems as low as you can, then remove the landscape fabric before applying the compost. You can spread compost directly on top of spent salad greens or mow the greens first if they are bushy. Optionally, cover the entire plot with silage tarp to hold the ground over winter and to create a tidier look. In March, you can pull back the tarp and start farming.

2. *Revisit your customers.* Ask them how you could improve in the next season. Revisit the three questions: What do you want? When do you want it? How much do you want? Deepen your relationship with them. The goal here isn't to find ways to add to your work but to find ways to essentialize it, to do less but better next season by more tightly aligning what you do with what your customer wants. A deep, open relationship with customers is the key to farming successfully on a hyperlocal scale.

   Now is also a good time to essentialize crops and customers. You want to skim from the surface: Which 20 percent of your crops performed and sold best? Which 20 percent of your customers generated the most income? Focus on those, the cream at the top, so to speak. Reduce your efforts on (or eliminate) the remaining 80 percent.

3. *Plan for next year: Essentialize and simplify fieldwork.* Take a minute to ponder your work, applying Pareto's analysis. Ask: Which tools got used? Which ones didn't? Purge anything that you didn't use. Remember the rule: Farm with as few tools as possible. Less is probably better.

   Similarly, which steps in the field took too long? Which methods seemed overly complicated? As I said earlier, every process can be simplified; it is just a matter of discovering how. I suggest you pick two or three methods to improve. We've found that analyzing our work like this—choosing a few new things to try doing differently next time around—motivates us to get back out in the garden for another year.

## Conclusion: Small Is Still Beautiful

As of this writing, Rachel and I have completed our fourth growing season using the tools and methods shared in this book. I can attest that this plan is achievable.

The key is to embrace small as an advantage, not a drawback. We now realize that by downsizing and simplifying, Rachel and I gave our farm a tremendous edge. We now drive very little to deliver our food compared to before. We have time to innovate and to work ahead instead of constantly playing catch-up. With the help of a very small crew, we can tend everything we plant because the farm is more human-scaled and manageable.

By following the get-small principles in this book, we have been able to meet our goals: to work fewer than 35 hours per week, to grow on ⅓ acre or less, to sell all of our food within 1½ miles of the farm.

Thinking lean hasn't taken the joy out of our work; it has freed up time to spend with our kids. Further, smallness—doing less—has helped us to be more relaxed and grounded, to be more aware and mindful and less distracted. There is a peaceful predictability to our work. Instead of just rushing to get it all done, we are now more apt to soak it in—to really enjoy the tactile, sensory experience of tending plants.

And getting small has increased our resilience: We waste less every season and use fewer materials, supplies, and tools every year to get our work done. Unexpectedly, our profits haven't suffered. Though we farm on a much smaller footprint, we earn as much as before.

The kids' garden at Clay Bottom Farm.

The get-small principles and practices in this book can transform your farm and life, too. All you need to get started is a list of your deepest values and a commitment to live and work on a human scale. Smallness—not gigantism—makes it possible to access ancient wisdom in our daily living, to do better with less.

In fact, think of the new world we could create if, instead of constantly doing more, we all did a little less but in a more efficient and conscientious way. In agriculture, we need farmers of all types to pivot toward smaller operations—alongside legions of new farmers willing to turn yards into gardens and vacant lots into vegetables. Or even just to put a plant in a windowsill and tend it well. At whatever level, wherever you are, that is the place to start.

By starting small, we inspire each other. Or as Schumacher put it:

> If one could make visible the possibility of alternatives, viable alternatives, make a viable future already visible in the present, no matter on how small a scale . . . then at least there is something, and if that something fits it will be taken. Suddenly, there will be demand. . . . If little people can do their own thing again, then perhaps they can do something to defend themselves against the overbearing, big ones.
>
> So I never feel discouraged. I can't myself raise the winds that might blow us, or this ship, into a better world. But I can at least put up the sail so that when the wind comes, I can catch it.[3]

I hope this book inspires you to put up your sail, to join the micro farm movement, to plant a seed.

# Minimalist Growing Tips for Seven More Crops

1. Cilantro
2. Cucumbers
3. Carrots
4. Basil
5. Sugar snap peas
6. Greenhouse figs
7. Hemp

Here are growing tips for seven more crops on a tiny scale. The first five are high-margin crops that can fit easily into the plan I shared in chapter 10. The final two, greenhouse figs and hemp, are more involved and probably lower-margin. They might not fit as well in the plan, at least in the first year, but they are still a delight to grow, with potential for profit in the right markets.

# Cilantro

*Variety:* Calypso

*Seeding dates:* March 1, May 1, July 1. With its long harvest window, only three plantings per year are required to supply cilantro from May through late fall.

*Final harvest:* October or November (later if in a hoophouse)

*Starting method:* Place 4 to 6 seeds per cell in 6″ paper chains or 128 plug flats

*Optimum germination temperature:* 70°F (21°C)

*Spacing:* 10″ between rows for four rows in a bed; 6″ between plants in-row. We grow cilantro from transplants, using the paperpot system. We use the 4mm top plate on the two-plate gravity seeder to dispense 4 to 6 seeds per cell, filling cells multiple times if needed. Cilantro is ready to transplant in about five weeks. It can be grown equally well from 128-cell plug flats or direct-seeded into weed-free beds.

**A.** Cilantro grows well in clumps like this and will bush out. **B and C.** To avoid covering small plants with the paperpot tool, remove the front plates. This will leave the paper chains in a shallow trench that can easily be filled in by hand or with a hoe.

Cilantro can be harvested multiple times—the trick is to cut often, forestalling flower formation. We harvest cilantro with the greens harvester. In summer, the best time to harvest cilantro is in the late evening when the leaves are dry and cool. We sell it in bulk bags to restaurants.

Three power crops (left to right): lettuce for spring mix, basil, and cilantro. With these crops, a high dollar value can be grown in a small space. Each of these crops will be harvested multiple times.

## Transplanting Leaned Up: The Two-Chains-per-Bed System

We grow basil, cilantro, head lettuce, spinach, turnips, and green onions with the Japanese-designed paperpot transplanter. While the tool is not for everyone, we wouldn't think of farming without it because it saves so much time and effort.

In fact, we designed our new farm around the transplanter. We chose a bed length of 75′ and a width of 42″ because that size comfortably fits two 6″ paperpot chains, at four rows per bed. (Each chain supplies two rows.) This is ideal spacing for head lettuce, cilantro, spinach, and many other crops. This system has removed an enormous amount of complexity from our farming. Choosing bed lengths to fit the paperpots also saves costs because the entirety of each chain gets used.

The paper chains that hold seedlings are the backbone of the system. These chains are spread out onto a frame, fit into a tray, and are filled with potting mix and seeds. Here are three tips for using paper chains:

*Tip #1:* When filling paper chains, use your fingers to press potting mix into the outer rows. Potting mix packs more densely in the middle part of the chains than around the edges.

*Tip #2:* Keep potting mix on the dry side. It should have some moisture in it so that it wicks water, but drier potting mix fills the cells more evenly.

*Tip #3:* Brush off the potting mix with a soft bristle brush, such as a hand dust brush, until you can see the top edge of the paper chain. This will make it easier to accurately use the dibbler tool and to fill the cells with seeds.

My book *The Lean Farm Guide to Growing Vegetables* explains how to use the tool in greater detail.

We designed our farm around the paperpot transplanter. Two 6″ paperpot chains fill one 75′-long bed. This spacing was used for the lettuce in this photo. Photo courtesy of Adam Derstine.

We start all lettuce heads for spring mix using this two-plate gravity seeder from Small Farm Works (www. smallfarmworks .com).

# Cucumbers

*Variety:* Corinto

*Seeding dates:* Seed on May 10 for transplanting June 1. Seed on June 1 for transplanting June 20. Seed on June 20 for transplanting July 10. This provides an even harvest over a long season.

*Starting method:* Place 1 seed per cell into 32-cell plug flats

*Optimum germination temperature:* 85°F (29°C)

*Spacing:* One row per bed, 18″ apart in-row spacing

*Trellising method:* We use Hortonova as a trellis net for cucumbers, with no pruning. This is much simpler and just as effective as training cucumbers to grow up twine, which is the more traditional approach. A tip I learned from Tiny Giant Farm: At the end of the season, remove cucumber vines from the trellis nets *before* they dry. This is faster than pulling dried vines from the nets, and this way you can reuse Hortonova nets for many seasons. Soak the nets between crops in a solution of OxiDate.

Helping young cucumbers vines to find the trellis netting. After initially helping plants find the net, we leave cucumbers to grow on their own. Photo courtesy of Caleb Mast Photography.

To protect cukes from cucumber beetles, we cover plants with row cover until the flowers appear, after which point plants are typically strong enough to resist the beetles.

**A.** For successive harvests, we seed cucumbers every three weeks. **B.** Notice here how we created a horizontal scaffold with twine that runs the length of the greenhouse. The rows of twine are 12″ apart. Once plants reach the greenhouse cross-ties, we push the tips over with a broom onto the twine, as with cherry tomatoes. Thus, the plants grow in an L shape. They grow equally well in a T shape if room allows. **C.** Here I am harvesting Corinto cucumbers that were trained to grow in an L shape.

# Carrots

*Varieties:* Yaya (spring), Bolero (fall)

*Seeding dates:* Anytime from March to May for summer harvest; June 20 for fall and winter harvest

*Starting method:* Direct-seed using the Jang seeder with roller F-12, brush low, and sprockets set at 13 front/11 rear. Use raw seed only (pelleted seeds offer slower and less consistent germination, in our experience).

*Optimum germination temperature:* 80°F (27°C), but they also germinate well at lower temperatures

*Spacing:* 8″ between rows (four rows per 42″-wide bed)

*Harvesting:* We harvest carrots with digging forks. If the ground is dry, we water for a few hours before harvest to make it easier to dislodge the roots.

*Washing:* We wash spring carrots with a hose, leaving tops on. We remove the tops of fall carrots and wash them with our homemade carrot washer. The carrot washer is made out of a mesh benchtop held in the shape of a barrel with two bicycle tire rims. The barrel is turned with the help of casters that fit into the tire rims (see also page 252). For the cleanest carrots, soak them in buckets for several minutes before hosing them.

**A.** For long carrots, loosen the ground before seeding. Here workshop participants help loosen this bed with a wheel hoe and Hoss sweeps. The E-Ox electric wheel hoe can also be used for this task. This is a faster method than broadforking. **B.** Seed carrots into moist soil, water well, and cover with three layers of Agribon AG-50 row cover or with landscape fabric until germination. The soil must stay moist for the entire germination period. **C.** Harvesting carrots with a digging fork. The variety here is Bolero. Note that we try to disturb the soil as little as possible during harvest in order to keep the 4″ layer of deep mulch intact. **D.** Washing carrots in a farm-built carrot washer.

# Basil

*Variety:* Prospera Italian Large Leaf DMR

*Seeding dates:* May 1 for June 1 transplanting in a greenhouse

*Starting method:* Seed 4 to 6 seeds per cell in 6″ paperpot chains or 128-cell plug flats. Due to their small size, this task is best done by hand.

*Optimum germination temperature:* 70°F (21°C)

*Spacing:* 24″ between rows, about two rows per bed; 6″ apart in-row spacing

*Harvesting:* We begin harvesting basil when it is about 10″ tall, and we cut it once per week all season. Harvest basil in the late evening or in the morning after the dew evaporates. After harvest, store leaves between 50 to 60°F (10–16°C) (note that this is warmer than a typical refrigerator). If you sell in plastic bags, a paper towel in the bag can help absorb moisture and prolong shelf life.

We transplant basil using 6″ paperpot chains, with 4 to 6 seeds per cell, as with cilantro. They also grow well in clusters, using 128-cell plug flats. Photo courtesy of Caleb Mast Photography.

Basil is a simple crop to grow, with a high yield per square foot.

# Sugar Snap Peas

*Variety:* Super Sugar Snap

*Seeding date:* April 15

*Starting method:* Soak seeds in water the night before seeding. Create a 6"-wide trench, 1" deep, with a shovel. Hand-seed 24 seeds per foot, spreading seeds out along the bottom of the trench, then cover with ½" of soil. Cover with row cover to keep soil moist until germination.

*Optimum germination temperature:* 70°F (21°C)

*Spacing:* 1 row per bed

**A.** We soaked these peas the night before planting them. **B.** Radishes are easily grown underneath spring peas. Note that the peas are trellised with a stake-and-weave system, as with slicing tomatoes (see page 212). **C.** Our biggest challenge growing peas: Late-spring heat can stress the plants just when the peas are ready to harvest. We mitigate the problem by building a temporary shade house.

# Greenhouse Figs

*Varieties:* LSU gold, LSU purple, Italian honey, and an unknown variety sourced from Jubilee Partners in Comer, Georgia

*Planting date:* Early spring

*Spacing:* 1 row per bed, with trees 8′ apart

*Management:* Train a main trunk to grow horizontally along a wire, about 18″ off the ground. Cut the growing tip of this trunk when it reaches 8′ long. (This trunk can also be grown in a T shape.) Every 12″ along the horizontal truck, train two leaders to grow up the twine, as with heritage tomatoes. These leading branches should grow away from one another, in the shape of a V, 24″ apart.

**A.** Figs are easily grown in northern greenhouses. Our plants have survived multiple winters without supplemental heat. **B.** We grow figs using a Japanese trellising method that involves training plants to one horizontal trunk, about 18″ off the ground. Each horizontal trunk is allowed to grow 8′ long. We support this trunk with ½″ metal conduit. Here, strawberries were planted as an understory crop. **C.** Fig tree after yearly pruning. After leaves fall, we prune the vertical branches to 2″ to 3″ a year, allowing two vertical shoots per foot to grow from the horizontal main branch.

Figs growing with an understory crop of kale. Notice that figs are trained with vertical shoots, about two shoots per foot. We grow the figs mostly for our own use. They may become a cash crop in the future, but we have not compared their yields per square foot to other greenhouse crops.

Vertical shoots, which support fruit production, grow from the main horizontal branch.

241

# Hemp

In 2019, we grew 100 hemp plants for CBD tinctures that we sell locally. We used the deep-mulch method, and we spaced plants 4′ apart. We grew them into landscape fabric, with drip tape underneath, the same approach that we use for tomatoes (see page 202). Compost was our only fertility source.

Hemp is a specialty crop, requiring certification to grow. It might not be the most lucrative crop to start with or the easiest to grow, but in our experience, it can fit easily within a diversified market garden setup. On a tiny-scale farm, I recommend growing hemp bred for high-CBD (or high-CBG) production instead of industrial or fiber hemp, which requires more land to be economically viable. See our farm website for an online course specifically about growing hemp on a microscale.

We started hemp by seed, using the same process as for tomatoes.

Here I am standing in front of hemp plants in the greenhouse. The biggest challenge with growing hemp in a greenhouse is bud rot. To mitigate the problem, we used greenhouse fans and grew the plants along an open sidewall for maximum ventilation.

Hemp next to carrots. In the field, we grew hemp alongside other crops. We grew it into landscape fabric and trellised it with oak stakes and twine, as with slicing tomatoes. We harvested in September.

Hemp plants must be dried in controlled conditions. Here we are drying hemp in our barn-house in a room with fans and dehumidifiers. Notice the use of cattle panels, hung from the ceiling, to support the plants. After the plants dried, we stripped the leaves and buds and sent them to a processor.

 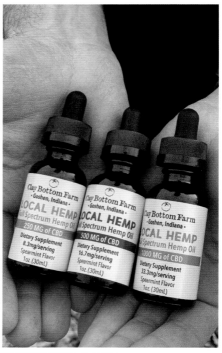

Dried leaves and buds were put into sling bags and weighed with hanging scales.

A processor extracted crude oil from our leaves and buds and processed it into these tinctures.

# Recommended Reading

## BOOKS ON MINIMALISM AND SIMPLE LIVING THAT SHAPED IDEAS IN THIS BOOK

*The Abundance of Less: Lessons in Simple Living from Rural Japan* by Andy Couturier (North Atlantic Books, 2017).

*Autobiography: The Story of My Experiments with Truth* by Mohandas K. Gandhi (Dover Publications, 1983).

*Big House, Little House, Back House, Barn: The Connected Farm Buildings of New England* by Thomas C. Hubka (University Press of New England, 1984).

*E. F. Schumacher, His Life and Thought* by Barbara Wood (Harper & Row, 1984).

*Essentialism: The Disciplined Pursuit of Less* by Greg McKeown (Crown Business, 2014).

*George Washington Carver: A Life* by Christina Vella (Louisiana State University Press, 2015).

*How to Eat* by Thich Nhat Hanh (Parallax Press, 2014).

*Just Enough: Lessons in Living Green from Traditional Japan* by Azby Brown (Tuttle Publishing, 2013).

*Less Is More: How Degrowth Will Save the World* by Jason Hickel (William Heinemann, 2020).

*Small Is Beautiful: Economics as If People Mattered* by E. F. Schumacher (Harper & Row, 1973).

*Subtract: The Untapped Science of Less* by Leidy Klotz (Flatiron Books, 2021).

*Ten Principles for Good Design: Dieter Rams: The Jorrit Maan Collection*, edited by Cees W. de Jong (Prestel Verlag, 2021).

## FOR MORE ON THE JAPANESE *HAN-NŌ, HAN-X* (HALF-AGRICULTURE, HALF-X) MOVEMENT

*Chikyu seikatsu: Gaia jidai no raifu paradaimu = Our true nature* by Jun
　　Hoshikawa (Tokyo: Tokuma Shoten, 1990) (in Japanese only).
"Half Farmer, Half X" by Institute for Studies in Happiness, Economy and
　　Society, 2011, https://www.ishes.org/en/cases/2011/cas_id000132.html.
"The Han-No Han-X Movement: Half-Agriculture, Half-Anything" by
　　Megumu Ogata and Nina Fallenbaum, *Slow Japan*, https://slowjapan
　　.wordpress.com/lifestyle.

## FOR MORE ON LEAN IN DEVELOPMENT WORK

Lean Toolkit for Market Systems Development, software available at
　　https://winrock.org/document/winrocks-lean-toolkit-for-market
　　-systems-development. This toolkit was developed by Charity Hanif,
　　Peter Sailing, Jessica Kelly, and myself and is published by Winrock
　　International.

## SCIENTIFIC STUDIES ON THE USE OF TARPS IN NO-TILL SYSTEMS

Grace Smith, Sonja Birthisel, and Eric R. Gallandt, "Comparing Solariza-
　　tion & Occultation," The University of Maine Weed Ecology and
　　Management, June 2017, https://umaine.edu/weedecology/2017
　　/06/01/comparing-solarization-occultation.
Natalie Hoidal, "Using the Sun to Kill Weeds and Prepare Garden Plots,"
　　University of Minnesota Extension, https://extension.umn.edu
　　/planting-and-growing-guides/solarization-occultation.

Sketches of Clay Bottom Farm barn-house. Courtesy of Gregory Lehman.

# Notes

## INTRODUCTION

1. Along the way, we also applied concepts from Taiichi Ohno, a manager at Toyota, and from *Lean Thinking: Banish Waste and Create Wealth in Your Corporation*, by James P. Womack and Daniel T. Jones (New York: Simon & Schuster, 1996).

2. Hannah Ritchie, Pablo Rosado, and Max Roser, "Fossil Fuels," Our World in Data, accessed October 10, 2022, https://ourworldindata.org/fossil-fuels.

3. E. F. Schumacher, *Small Is Beautiful: Economics as If People Mattered* (New York: Harper & Row, 1973), 74.

4. Schumacher, *Small Is Beautiful,* 34.

5. USDA Economic Research Service, "Climate Change," USDA-ERS, accessed December 22, 2022, https://www.ers.usda.gov/topics/natural-resources-environment/climate-change.

6. Gregory A. Baker et al., "On-Farm Food Loss in Northern and Central California: Results of Field Survey Measurements," *Resources, Conservation and Recycling* 149 (October 2019): 541–49, https://doi.org/10.1016/j.resconrec.2019.03.022.

7. David Yaffe-Bellany and Michael Corkery, "Dumped Milk, Smashed Eggs, Plowed Vegetables: Food Waste of the Pandemic," *New York Times*, April 11, 2020, accessed October 10, 2022, https://www.nytimes.com/2020/04/11/business/coronavirus-destroying-food.html.

8. Oleksandr Tarassevych, "Households Agricultural Production," USDA Foreign Agriculture Service, May 23, 2022, accessed February 27, 2023, https://apps.fas.usda.gov/newgainapi/api/Report/DownloadReportByFileName?fileName=Households%20Agricultural%20Production_Kyiv_Ukraine_UP2022-0030; Natasha Foote, "Small Farmers: The Unsung Heroes of the Ukraine War," *Euractiv*, April 20, 2022, accessed December 27, 2022, https://www.euractiv.com/section/agriculture-food/news/small-farmers-the-unsung-heroes-of-the-ukraine-war.

9. Bill Kiernan, "Grass Fed versus Corn Fed: You Are What Your Food Eats," *Global AgInvesting*, July 16, 2012, accessed November 2, 2022,

https://www.globalaginvesting.com/grass-fed-versus-corn-fed
-you-are-what-your-food-eats.

10. Vandana Shiva, *Who Really Feeds the World?: The Failures of Agribusiness and the Promise of Agroecology* (Berkeley, CA: North Atlantic Books, 2016), xix.

11. Schumacher, *Small Is Beautiful*, 39.

## CHAPTER 1: LEVERAGE CONSTRAINT

1. Barbara Wood, *E .F. Schumacher, His Life and Thought* (New York: Harper & Row, 1984), 244.

2. Thich Nhat Hanh, "The Art of Living," in *Mindfulness and Meaningful Work: Explorations in Right Livelihood*, ed. Claude Whitmyer (Berkeley, CA: Parallax Press, 1994), 245.

3. E. F. Schumacher, *Small Is Beautiful: Economics as If People Mattered* (New York: Harper & Row, 1973), 148.

4. Schumacher, *Small Is Beautiful*, 57.

## CHAPTER 2: BUILD *JUST ENOUGH*

1. E. F. Schumacher, *Small Is Beautiful: Economics as If People Mattered* (New York: Harper & Row, 1973), 64.

2. Tim Huber, "Russian Troops Occupy Region of Mennonite Ministry," *Anabaptist World*, February 28, 2022, accessed February 13, 2023, https://anabaptistworld.org/russian-troops-occupy-region-of
-mennonite-ministry.

3. A good resource on connected farms is Thomas C. Hubka's *Big House, Little House, Back House, Barn: The Connected Farm Buildings of New England* (Lebanon, NH: University Press of New England, 2004).

4. Hubka, *Big House, Little House, Back House, Barn*, 199.

5. Mark Dziersk, "In Tribute to Mr. Rams," *Fast Company*, February 12, 2012, accessed December 5, 2022, https://www.fastcompany.com
/1815761/tribute-mr-rams.

## CHAPTER 3: ESSENTIALIZE: THE PARETO PRINCIPLE

1. Early in his career, Juran called the 80 percent the "trivial many." He later changed his description of the 80 percent to the "useful many" to reflect his conviction that the 80 percent, in some cases, shouldn't always be discarded.

## CHAPTER 4: SIMPLIFY FIELDWORK

1. S. J. Scott, *Habit Stacking: 97 Small Life Changes That Take 5 Minutes or Less* (Pasadena, CA: Oldtown Publishing, 2017).

2. John Shook, "How Standardized Work Integrates People with Process," Lean Enterprise Institute, accessed December 27, 2022, https://www.lean.org/the-lean-post/articles/how-standardized-work-integrates-people-with-process.

3. V. van Vliet, "Taiichi Ohno Biography and Quotes," Toolshero, last modified December 7, 2022, accessed December 27, 2022, https://www.toolshero.com/toolsheroes/taiichi-ohno.

4. Michael Lewis, "Obama's Way," *Vanity Fair*, September 11, 2012, accessed February 24, 2023, https://www.vanityfair.com/news/2012/10/michael-lewis-profile-barack-obama.

5. Barbara Wood, *E. F. Schumacher, His Life and Thought* (New York: Harper & Row, 1984), 244–45.

6. E. F. Schumacher, *Small Is Beautiful: Economics as If People Mattered* (New York: Harper & Row, 1973), 154.

7. Joseph R. Heckman, Daniel Kluchinski, and Donn A. Derr, "Plant Nutrients in Municipal Leaves," Rutgers Cooperative Research and Extension, Fact Sheet FS824, 2004, accessed November 16, 2022, https://sustainable-farming.rutgers.edu/wp-content/uploads/2014/09/Municipal_Leaves_Plant_Nutrients_Available_FS824_1998.pdf.

8. Micheal Snyder, "What Do Tree Roots Do in Winter?," *Northern Woodlands*, December 1, 2007, accessed November 16, 2022, https://northernwoodlands.org/articles/article/what_do_tree_roots_do_in_winter.

9. George Washington Carver, "How to Build Up Worn Out Soils," Tuskegee Normal and Industrial Institute, Bulletin 6, 1905, accessed December 5, 2022, https://archive.org/details/CAT31355455/page/n1/mode/2up.

10. Phil Edwards, "George Washington Carver Cared about Sustainable Farming Before It Was Cool," *Vox*, July 13, 2015, accessed December 5, 2022, https://www.vox.com/2015/7/13/8948477/george-washington-carver.

11. Linda O. McMurry, *George Washington Carver, Scientist and Symbol* (New York: Oxford University Press, 1982), 130.

12. E. F. Schumacher, *Small Is Beautiful: Economics as If People Mattered* (New York: Harper & Row, 1973), 154.

## CHAPTER 5: LOCALIZE: THE PRACTICE OF *SWADESHI*

1. Nergish Sunavala, "Gandhi March: In Search of the Mahatma, from Parel to Girgaon," *Times of India*, September 30, 2018, accessed February 7, 2023, http://toi.in/B0ZLaZ66/a26gk.

2. Mahatma Gandhi, *Third Class in Indian Railways* (Lahore, Pakistan: Gandhi Publications League, 1917), 11.

3. "Localize," Dictionary.com, accessed November 28, 2022, https://www.dictionary.com/browse/localize.

4. "Localize," Oxford Languages, Google, accessed November 28, 2022, https://www.google.com/search?q=localize+definition.

5. Simone Panter-Brick, *Gandhi and Nationalism: The Path to Indian Independence* (New York: I.B. Tauris, 2012).

6. "Gandhi's Views on Swadeshi/Khadi," Sevagram Ashram, accessed April 17, 2023, https://www.gandhiashramsevagram.org/swadeshi/definition-of-swadeshi.php.

7. Gandhi, *Third Class in Indian Railways*, 16.

8. E. F. Schumacher, *Small Is Beautiful: Economics as If People Mattered* (New York: Harper & Row, 1973), 6.

9. adrienne maree brown, *Emergent Strategy: Shaping Change, Changing Worlds* (Chico, CA: AK Press, 2017), 87.

10. Thich Nhat Hanh, *Awakening of the Heart: Essential Buddhist Sutras and Commentaries* (Berkeley, CA: Parallax Press, 2011), 413–14.

11. "Clans," Pokagon Band of Potawatomi, accessed November 28, 2022, https://www.pokagonband-nsn.gov/our-culture/clans.

12. Shannon Martin, *Start with Hello: (And Other Simple Ways to Live as Neighbor* (Ada, MI: Revell, 2022).

13. E. F. Schumacher, *This I Believe: And Other Essays* (Cambridge, UK: Green Books, 1998), 72–73.

## CHAPTER 6: NO-TILL FOR MICRO FARMS: THE DEEP-MULCH METHOD

1. Barbara Wood, *E. F. Schumacher, His Life and Thought* (New York: Harper & Row, 1984), 369.

2. "Monitoring Compost pH," Cornell Waste Management Institute, 1996, accessed December 7, 2022, https://compost.css.cornell.edu/monitor/monitorph.html.

3. Steven Wisbaum, "Low-Input Composting," CV Compost, revised September 2021, accessed December 5, 2022, https://www.cvcompost .com/low-input-composting.
4. Wisbaum, "Low-Input Composting."

## CHAPTER 7: TWO-STEP BED FLIPPING: A METHOD FOR INCREASING YOUR PRODUCTION ON A SMALL FOOTPRINT

1. David A. Zuberer, "Soil Microbiology FAQs," Texas A&M University, accessed December 13, 2022, http://organiclifestyles.tamu.edu/soil /microbeindex.html.
2. Elizabeth Bernhardt and Ted Swiecki, "Using Heat to Eradicate Soil-Borne Plant Pathogens from Nursery Potting Media ("Soil Sterilization")," *Phytosphere Research*, updated March 31, 2021, accessed December 13, 2022, http://phytosphere.com/soilphytophthora /soilsterilization.htm.
3. James J. Hoorman and Rafiq Islam, "Understanding Soil Microbes and Nutrient Recyling," *Ohioline*, Ohio State University Extension, September 7, 2010, accessed December 12, 2022, https://ohioline.osu .edu/factsheet/SAG-16.

## CHAPTER 9: *LESS BUT BETTER* INFRASTRUCTURE

1. Roger Fisher and William Ury, *Getting to Yes: Negotiating Agreement Without Giving In* (New York: Penguin Books, 1981).

## CHAPTER 10: GETTING STARTED: A PLAN FOR SELLING $20,000 FROM YOUR BACKYARD

1. Kimberly Hughes, "Half-Farming, Half-Anything: Japan's Rural Lifestyle Revolution" *Japan Times*, January 21, 2023.
2. Barbara Wood, *E. F. Schumacher, His Life and Thought* (New York: Harper & Row, 1984), 369.
3. E. F. Schumacher, *Good Work* (New York: Harper Colophon Books, 1979), 64–65.

# Index

# About the Author

Ben Hartman is the author of *The Lean Farm* (Chelsea Green, 2016), winner of the prestigious Shingo Institute Research and Professional Publication Award. In 2017, Ben was named one of fifty emerging green leaders in the United States by *Grist* and published a companion guide to *The Lean Farm* titled *The Lean Farm Guide to Growing Vegetables*. Ben and his wife, Rachel Hershberger, own and operate Clay Bottom Farm in Goshen, Indiana, where they make their living growing specialty crops on one-third acre. Ben has developed an online course in market gardening on a micro scale, which can be found at claybottomfarm.com, where you can also read more about the farm.